Astronomers' Universe

Series Editor
Martin Beech, Campion College
The University of Regina
Regina, Canada

The Astronomers' Universe series attracts scientifically curious readers with a passion for astronomy and its related fields. In this series, you will venture beyond the basics to gain a deeper understanding of the cosmos—all from the comfort of your chair.

Our books cover any and all topics related to the scientific study of the Universe and our place in it, exploring discoveries and theories in areas ranging from cosmology and astrophysics to planetary science and astrobiology.

This series bridges the gap between very basic popular science books and higher-level textbooks, providing rigorous, yet digestible forays for the intrepid lay reader. It goes beyond a beginner's level, introducing you to more complex concepts that will expand your knowledge of the cosmos. The books are written in a didactic and descriptive style, including basic mathematics where necessary.

Gianpaolo Bellini

How the Sun and the Stars Shine

A Single Experiment Solves a Millennia-Old Question of Humanity

Gianpaolo Bellini
Physics Department
Università degli Studi
Milano, Italy

ISSN 1614-659X ISSN 2197-6651 (electronic)
Astronomers' Universe
ISBN 978-3-031-88622-5 ISBN 978-3-031-88623-2 (eBook)
https://doi.org/10.1007/978-3-031-88623-2

© The Editor(s) (if applicable) and The Author(s), under exclusive license to Springer Nature Switzerland AG 2025

This work is subject to copyright. All rights are solely and exclusively licensed by the Publisher, whether the whole or part of the material is concerned, specifically the rights of translation, reprinting, reuse of illustrations, recitation, broadcasting, reproduction on microfilms or in any other physical way, and transmission or information storage and retrieval, electronic adaptation, computer software, or by similar or dissimilar methodology now known or hereafter developed.
The use of general descriptive names, registered names, trademarks, service marks, etc. in this publication does not imply, even in the absence of a specific statement, that such names are exempt from the relevant protective laws and regulations and therefore free for general use.
The publisher, the authors and the editors are safe to assume that the advice and information in this book are believed to be true and accurate at the date of publication. Neither the publisher nor the authors or the editors give a warranty, expressed or implied, with respect to the material contained herein or for any errors or omissions that may have been made. The publisher remains neutral with regard to jurisdictional claims in published maps and institutional affiliations.

This Springer imprint is published by the registered company Springer Nature Switzerland AG
The registered company address is: Gewerbestrasse 11, 6330 Cham, Switzerland

If disposing of this product, please recycle the paper.

To Nice, my invaluable wife

Foreword

How the Sun shines is a very important question for us living on the Earth. We live on Earth thanks to the light radiated from the Sun. It has long been known that observing neutrinos from the Sun can reveal how the Sun shines. However, this is experimentally very difficult, because the neutrinos pass through everything easily.

Gianpaolo Bellini is a distinguished physicist who led a solar neutrino experiment, called Borexino, for more than 30 years, and showed how the Sun shines. As you will see reading this book, this was a remarkable achievement made possible by the collaboration of many scientists who contributed to this success through their diverse expertise and wisdom.

I hope that, through this book, many people will learn how great scientific achievements are made.

Nobel Laureate, 2015

Takaaki Kajita
Special University Professor at the University of Tokyo

Humanity's age-old question about the nature and workings of the Sun and stars has finally been answered in recent years through a groundbreaking experiment featuring a one-of-a-kind detector.

Competing Interests The author has no competing interests to declare that are relevant to the content of this manuscript.

Contents

1 Introduction and Guidance for Readers   1
  1.1 Introduction   1
  1.2 Guidance for the Reader   2

2 The Context   5
  2.1 A Scientific Experiment: What Is It?   5
  2.2 The Sky Above Us   7
  2.3 A Missing Piece   25
  2.4 Neutrinos and Solar Neutrinos, Formidable Probes   27
  Annex 2.1: Nuclear Fusion Reactions in the pp Chain operating in the Sun   31
  Annex 2.2: The CNO Cycle   32

3 The Genesis of the Experiment   33
  3.1 The Solar Neutrino Problem   34
  3.2 Borex and Borexino   37
  Annex 3.1   41

4 The Battle for Unprecedented Radiopurity   43
  4.1 The Start Up   43
  4.2 The "Mission Impossible" Truly Begins   45
  4.3 Are We Capable of Winning the Battle of Radiopurity?   47
  4.4 The Moment of Truth   53
  4.5 The Counting Test Facility Is Being Upgraded for Further Important Tests   56

5	**First Phase of Detector Construction**	59
	5.1 The Mission Impossible Begins	59
	5.2 A Long Task: The Construction of the Detector	64
	5.3 Two and More Dark Years	75
6	**Completion of Construction and Installation: The Detector Goes into Operation**	83
	6.1 The Resumption	83
	6.2 The Radiopurity	87
	6.3 The Set Up Within the Sphere	87
	6.4 Installation of Nylon Container for Scintillator: Inner Vessel—IV	94
	6.5 Inside the Water Tank	96
	6.6 What Happens to the Data Collected by the Photomultipliers?	98
	6.7 The Filling	100
	Annex 6.1	105
7	**What Powers the Sun?**	107
	7.1 The Detector Operates Effectively	107
	7.2 Flash on the Sun's Structure	110
	7.3 The Calibration	113
	7.4 A Second Radio-Purification	116
	7.5 We Discover What Happens in the Sun and How Its Energy Is Produced	117
	Annex 7.1	121
	Annex 7.2	122
	Annex 7.3	123
	Annex 7.4	124
8	**Sun Stability and Earth Orbit**	125
	8.1 The Sun Stability	125
	8.2 Calculating the Earth's Orbit Using Solar Neutrinos	126
	Annex 8.1	127
9	**And the Stars?**	129
	9.1 A Fourth Neutrino?	129
	9.2 How Do We Investigate Stars?	130
	9.3 The Sun Metallicity	135
	9.4 Annex 9.1	137

10	**Corollaries**	139
	10.1 Neutrino Oscillation	139
	10.2 Geoneutrinos	140
	Annex 10.1	143
11	**Farewell Borexino**	145
	11.1 Farewell Borexino	145
	11.2 The Borexino Legacy	146
Glossary		149

About the Author

Gianpaolo Bellini, emeritus professor of the University of Milan, emeritus scientist of the National Institute for Nuclear Physics, and experimental physicist in the fields of elementary particles and astroparticles, has performed and directed various experiments in the most important laboratories in the world such as CERN (Geneva, CH), IHEP (Protvino, Russia), Fermi National Laboratory (Batavia, USA), and Gran Sasso laboratories (Assergi, Italy). His discoveries concern resonances, coherent interactions on nuclei, and lifetime of particles with the charm quark, the mechanisms that make the Sun and stars shine. He has been rewarded with the prestigious international Bruno Pontecorvo Prize, the Enrico Fermi Prize, and the Wanda and Giuseppe Cocconi Prize of the European Physical Society addressed to the Borexino Collaboration. The Borexino experiment conceived and directed by him at the Gran Sasso laboratories produced results that have been nominated among the world's ten best breakthroughs in 2014 and 2020, while a celebratory stamp has been dedicated by the Italian post office. He is the author of about 210 papers in international journals and editor of 10 books on elementary particle physics and 4 books on popular science.

1

Introduction and Guidance for Readers

1.1 Introduction

The Sun and the stars have accompanied the lives of all human beings since the dawn of humanity. There is no human who has not been warmed by the Sun's rays or, perhaps, gazed at the starry sky at night. But I wonder how many have thought about how long the Sun will last in its current state—one of the countless conditions that make life on Earth possible—and how our star has so much energy to spread light and heat throughout the solar system. Perhaps, few people know that the light reaching us today was actually produced on the average 100,000 years ago, around the time when Neanderthals occupied much of Europe and western Asia, and Homo sapiens began their first migrations out of Africa. Similarly, some might have wondered how the stars of the Milky Way, so beautifully visible at night, may have enough energy to emit light that travels from tens to hundreds of thousands of years before reaching our Earth, an infinitesimal point in the vastness of the universe.

Astronomy and astrophysics have uncovered many things about how the universe around us was formed and how it works. However, when it came to understanding how the Sun and stars were capable of producing so much energy and therefore so much light, hypotheses were made, but there was no direct evidence of what actually happens inside these celestial bodies.

To understand the mechanisms that power the Sun and the stars, it is crucial to explore their interiors, as energy production occurs in the core of them. Given that the temperature inside them reaches millions of degrees, it is evident that no man-made object could function as a probe. However, there exist

© The Editor(s) (if applicable) and The Author(s), under exclusive license to Springer Nature Switzerland AG 2025
G. Bellini, *How the Sun and the Stars Shine*, Astronomers' Universe, https://doi.org/10.1007/978-3-031-88623-2_1

extraordinary natural probes capable of fulfilling this role: one of the fundamental building blocks of matter, known as *neutrino,* which can travel completely undisturbed through the matter that makes up the Sun and the stars.

In recent decades, an experiment installed in the Apennine mountain range in Italy, under 1400 m of rock, has been able to study the Sun's interior with the help of these natural probes and identify the mechanisms that make the Sun and stars shine. This experiment lasted 31 years and was both a great scientific and human adventure, as such an endeavor would not have been possible without a profound passion for science—no one would dedicate half a lifetime to conducting such an experiment otherwise.

I was one of the principal architects and responsible of this experiment, and I decided to publish this book to share the story of this adventure, which is not only scientific but also human. My aim is to help even readers who may not have a particular interest in, or connection to, scientific activities understand what it means to do science, how science progresses, and the countless small steps required to try to comprehend how the world around us works. Certainly, the results of this experiment represent a small step, a tiny piece in the vast puzzle of human knowledge. However, everything we know today is the result of the collective assembly of such pieces, particularly over the past century and a half. There is still so much to understand, and with each new discovery, we uncover more of what we have yet to grasp; however, I think it is fascinating for everyone to understand how all of this works and to better understand the world in which we are so deeply immersed.

I have tried to describe this experiment in the simplest way possible, so it can be understood by anyone, and I also wanted to present it as a human adventure that involves the lives of some people, and in any case, the work of many people.

I hope that the readers of this book, regardless of their background, will be able to understand the work carried out and the results achieved, and that, in the end, he will have gained something by learning a little more about the sky above us. I would like to recall the words of the great scientist Albert Einstein, who said that the universe around us is already a miracle, but an even greater miracle is that it is intelligible to humankind.

1.2　Guidance for the Reader

When discussing a scientific experiment, one cannot avoid describing and explaining scientific and technical aspects, all within the context of a human endeavor. In this book, I have used language that is as simple as possible for a general reader, and I have omitted some descriptions and figures typical of the scientific language regarding the experiment, which lies at the intersection of particle physics and astrophysics—what is currently referred to as astroparticle physics. I hope, therefore, that it will be understandable to everyone. However, I have included annexes, which are in-depth information for those with some scientific knowledge who are eager to delve deeper and understand more of what is explained in the text. These annexes are placed at the end of each chapter.

2

The Context

2.1 A Scientific Experiment: What Is It?

The central focus of this book is a scientific experiment that uncovered how the Sun and stars generate their light and energy. Before delving into the details of this subject, it may be helpful to address some fundamental questions about scientific experimentation: What is a scientific experiment? Why is it undertaken? Why choose one experiment over another? How does the idea for an experiment arises? What resources and preparations are needed to carry it out?

Scientists undertake experiments for various reasons. For example, a prior experiment might not have sufficiently explained the phenomena under investigation, necessitating further experiments to clarify unresolved aspects. Alternatively, a scientist may be driven by fascination with a topic, motivating deeper exploration. In my case, I pursued the experiment discussed here because I believed that understanding why and how the Sun and stars shine has been a profound and timeless question for humanity. I also felt it was essential to provide a definitive answer to this mystery.

Often, a scientist's curiosity is sparked by theoretical hypotheses or models that need experimental validation—the only way to determine if these ideas accurately reflect reality. Every hypothesis, model, or scientific theory must be tested experimentally. Without experimental confirmation, a hypothesis or theory cannot be considered truly scientific; it remains a mathematical model, a philosophical idea, or something similar. Take, for example, Galileo Galilei's hypothesis in the early seventeenth century: he proposed that all objects fall to Earth with the same acceleration, regardless of their shape or weight. Galileo tested this by dropping objects from the Leaning Tower of Pisa, but his results were complicated by air resistance. When it later became possible to conduct experiments in a vacuum—eliminating the friction of the atmosphere—it was observed that a feather and a lead ball fell with the same acceleration, confirming Galileo's hypothesis.

In some cases, direct experiments are not feasible, and scientists must rely on observations. Consider the motion of planets: models describing their movements have evolved over centuries. Initially, these models blended observation with philosophical ideas, such as the belief that celestial trajectories must follow perfect circles. Over time, increasingly precise observations led to a shift from the geocentric Ptolemaic system to the heliocentric model proposed by Copernicus. Kepler later refined this model by abandoning circular orbits in favor of elliptical ones, and Newton perfected it further by introducing the concept of universal gravitation and applying infinitesimal calculus to accurately calculate planetary motions. This progression—from observation to a correct understanding of planetary motion—was driven by observational data, as no experiments could be conducted on the planets themselves.

Another example is the Big Bang model, which hypothesizes that the Universe originated from the explosion of extremely dense, high-temperature matter. Since it is impossible to directly observe the Big Bang, this model is validated through observations of the Universe at increasingly distant times, corresponding to moments closer to the event itself. Recent satellite missions equipped with advanced detectors have captured radiation emitted by stars billions of years ago, providing evidence that supports the Big Bang model.

Once the idea for an experiment takes shape, the next step is to delve into its scientific aspects: designing an experiment that effectively tests the idea and developing a detector suited to the purpose. Simultaneously, it is crucial to secure support from funding agencies and/or scientific institutions. Their decision to back an experiment often depends on the proponent's reputation within the scientific community, built upon prior successes and achievements, his dedication to the scientific activities, and his ability to rally support from other research groups. A successful proposal often inspires interest—sometimes even enthusiasm—among researchers, fostering collaboration.

In many European countries, researchers are free to pursue topics that captivate them, driven solely by the scientific merits of the idea. In contrast, the process in the United States is somewhat different: researchers are often contracted to work on specific experiments, with their responsibilities clearly defined.

Embarking on a challenging experiment demands confidence, determination, and an intense enthusiasm for research. The journey is unpredictable—you may encounter unforeseen difficulties, requiring significant intellectual effort and persistence to overcome. Experiments often demand long hours and substantial sacrifices, but the rewards can be immense if the experiment succeeds and leads to groundbreaking scientific discoveries.

A theoretical physicist I know once summed up this passion well. When asked how long he had been working, he replied that he had never worked a day in his life—he was simply having fun, captivated by his studies.

Science progresses according to its own internal logic, driven by the quest to understand phenomena that remain unexplored or insufficiently explained. The decision to pursue a particular research path ultimately lies with the scientists themselves, especially in the realm of fundamental science, where the sole purpose is to unravel the mysteries of nature. This contrasts with applied science and technology, where choices are heavily influenced by economic, political considerations, wish for success.

2.2 The Sky Above Us

For the famous French cartoonists René Goscinny and Albert Uderzo, their indomitable heroes, Asterix and Obelix, have only one fear: *"que le ciel leur tombe sur la tête"* (that the sky will fall on their heads). The sky—this vast blanket of tiny lights hanging above us—has long been a spectacle that, until just a few decades ago, was clearly visible from our latitudes (an example in Fig. 2.1). However, due to light pollution, it is no longer something we can observe easily, unless we venture into the high mountains or the open sea, far from the coasts. This is a great loss. Yet, the sky does not only shine with stars; more importantly, it shines with the rays of the Sun, which play a decisive role in regulating our lives.

While Goscinny and Uderzo set their stories during the period of Gaul's conquest by Julius Caesar, around the first century BC, humanity's fascination with the sky began long before that, stretching back to the earliest days of mankind and certainly into prehistory.

Prehistory As early as the Paleolithic period (ranging from 2.7 to 2 million years ago up to about 10,000 years ago), when humans were primarily hunters and gatherers using tools made of chipped but unpolished stone (Paleolithic means "Old Stone Age"), evidence of systematic observation of the natural world can be found. Some of the earliest examples date back as far

Fig. 2.1 Starry sky with the Milky Way galaxy to which we belong, seen edge-on. The Milky Way is visible here as that white band in the sky. (*Source: Pixabay*)

as 40,000 years ago. In France, a bone fragment from an eagle's wing, dating to around 34,000 years ago, was discovered in the Dordogne region. Markings on this fragment are interpreted as lunar cycles, and their shape suggests the waxing and waning phases of the moon Fig. 2.2). Similar markings, thought to represent moon phases, have been found on a baboon bone from about 21,000 years ago near Lake Edward, on the border between Uganda and Congo. This bone features a series of symbols carved into three columns, which may represent a primitive lunar calendar. Further evidence comes from the caves of Lascaux in southwestern France, dating back to around 17,500 BC. Among the more than 6000 depictions of animals (such as horses, deer, cattle, bison, felines, birds, bears, and rhinoceroses), humans, and abstract symbols, there are what can be considered the earliest "star maps." These representations include stars that were particularly bright in the night sky, such as the Pleiades cluster. Some scholars believe that constellations were already recognized during the Paleolithic, and that by around 16,000 BC, a system of 25 constellations may have been established. Similar engravings have also been discovered in Mexico and Australia.

Our ancestors understood that as daylight faded, the night sky—adorned with a faithful and unchanging mantle of lights—would always appear. They observed the bright light of the Moon illuminating the dark, often dangerous

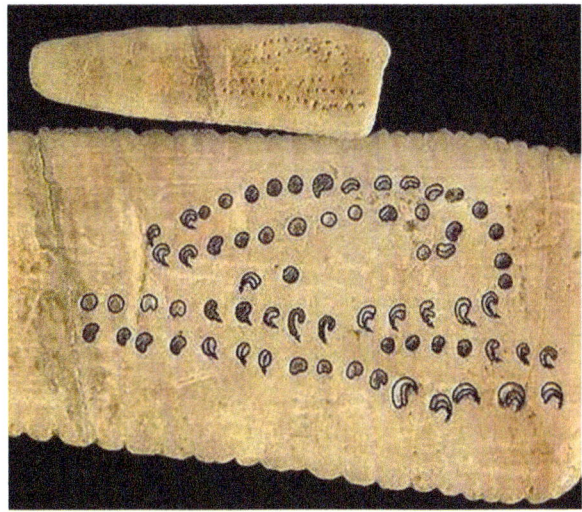

Fig. 2.2 Ancient bone artifacts with engraved symbols. The larger bone features a spiral pattern of circular and crescent-shaped carvings, while the smaller bone displays a series of linear dot engravings. The background is black, highlighting the intricate details of the carvings (34,000 years ago). (*Source: Anna Maragno-Università di Ferrara-scienza per tutti-INFN*)

landscape and recognized its recurring cycle. These early humans not only possessed language but also the ability to count, noting that the lunar cycle lasted approximately 29 days, roughly aligning with the female fertility cycle. As humans moved across the Earth, they encountered varying landscapes, vegetation, and fauna. Yet, the sky remained constant, reappearing each evening with subtle variations tied to the Sun's annual cycle. They attributed divine qualities to phenomena beyond their control, such as the constellations, rain, wind, and fire. Among these deities, the Sun God stood supreme, governing life through its yearly cycle.

During the Neolithic period (approximately 10,000–3500 BC), these celestial observations advanced significantly, leading to the codification of the zodiac constellations. The Sun's behavior was studied not only for religious and divinatory reasons but also for practical purposes. By this time, humans had achieved significant progress in domesticating animals such as sheep, goats, pigs, and oxen, as well as cultivating crops like cereals and legumes. Human settlements stretched from the Andes in South America to the Fertile Crescent, Anatolia, and other parts of the Middle East, as well as northern and central Africa. In this agro-pastoral economy, understanding the cycles of the year was crucial for activities such as sowing, harvesting, and breeding livestock.

One remarkable testament to early human knowledge of the Sun's behavior is the iconic structure of Stonehenge (Fig. 2.3). Built between 5000 and

Fig. 2.3 The Stonehenge stones (literally: hanging stone). (*Source: pixabay*)

4000 years ago on the Salisbury Plain in southern England, this arrangement of massive stone monoliths, some weighing tens of tons, features lintels raised 8 m above the ground. While likely serving religious purposes as a kind of temple, its most fascinating feature is its precise alignment with the Sun. The main axis of the structure aligns with the direction of the sunrise on the summer solstice. Secondary alignments correspond to the winter solstice and the two equinoxes—key intermediate points in the Sun's annual journey.

The people of this era clearly recognized the fundamental importance of the Sun God for their survival. It became a central figure in their spirituality, reflecting humanity's early understanding of the Sun's role in shaping life on Earth.

Another remarkable testament to Neolithic ingenuity is the Newgrange monument in Ireland (Fig. 2.4), which is believed to predate Stonehenge by approximately 500 years. Likely constructed as a tomb, Newgrange features an extraordinary alignment: at dawn on the winter solstice, sunlight enters through a narrow opening above the roof and travels down a 19-m-long corridor to illuminate the floor of the burial chamber.

The First Historical Civilizations The first historical civilizations made significant advancements in astronomy. The Babylonians and Egyptians (3000–2700 BC) achieved remarkable progress by developing monthly, annual, and seasonal calendars based on highly precise observations of the

Fig. 2.4 The Sun's rays penetrate Newgrange at dawn on the winter solstice. (*Source: pixabay*)

movements of Sun and Moon. As in prehistoric times, these advancements served both divinatory purposes—blending astronomy with astrology—and practical needs related to agriculture and daily human life. Slightly more recent contributions came from the Chinese and the Maya. The Babylonians demonstrated remarkable precision in their astronomical measurements of the Sun, stars, and planets. They observed that the Sun, over the course of a year, moves relative to the fixed stars along a path they already identified as the "ecliptic." These solar movements explained the changing lengths of day and night, while the Moon completed its cycle in roughly a month. The Babylonians also succeeded in linking the phases of the Moon to the Sun's position, enabling them to predict lunar and solar eclipses.

Similarly, the Egyptians exhibited advanced astronomical understanding. For example, the Giza pyramid's alignment along the north-south axis is accurate to within just five hundredths of a degree! These precise construction techniques were invaluable for navigators, who determined the direction of the North Pole by aligning two bright stars: Mizar in the Big Dipper and Kochab in the Little Dipper.

Both Babylonian and Egyptian astronomers attributed the events in the sky, as well as the origin of the cosmos, to divine forces. Consequently, they did not develop geometric or physical explanations for the birth of the universe.

In contrast, the Jewish civilization's worldview, as reflected in the Old Testament, was fundamentally different and distinct from those of the Babylonians and Egyptians.

The Greek Age The Greeks made some of the most significant contributions to astronomy by processing and refining observations from earlier civilizations (In Fig. 2.5 is depicted Raphael's fresco, which portrays the Greeks who initiated and developed the philosophy, the logic, and the science of ancient Greece).For the first time, they proposed that the Earth was a sphere rather than a flat object, and they introduced the concept of heliocentrism. The greatest advancements emerged from the schools of Miletus and Pythagoras. Thinkers from the Miletus school, such as Thales, Anaximander, and Anaxagoras, developed geometry and trigonometry to interpret the heavens. Thales used geometry to measure the Earth and demonstrated that the ratio

Fig. 2.5 Fresco by Raphael in the Vatican Loggias, dated to the 1520s. At the center of the composition, Plato and Aristotle are prominently featured at the top. In the lower section, the school of Pythagoras is depicted on the left, while a group of geometers and mathematicians is shown on the right. This iconic fresco, titled *The School of Athens* by Raphael, is a masterpiece of Renaissance art. (*Photo source: Wikipedia-The Vatican Museums-Vatican city*)

between the diameters and distances of the Moon and Sun was approximately 1:113. His student Anaximander proposed that the Sun emits light, while the Moon and Earth are opaque bodies, correctly interpreting the phenomena of eclipses.

A new era began with the Pythagorean school, which emphasized numerical calculations and believed that the universe was governed by numbers, proportions, symmetry, and harmony. This reverence for geometry and harmony led them to view circular and spherical shapes as the epitome of perfection. Consequently, they concluded that all celestial bodies were spherical and their movements were circular, eternal, and maintained by the harmony and intelligence of the cosmos. They also asserted that the Earth was spherical and followed a circular trajectory, like all other celestial bodies. However, their observations of planetary movements showed deviations from perfect circular orbits.

Philolaus, a pupil of Pythagoras, challenged the geocentric view. He argued that the Earth was too imperfect to occupy the center of the universe. Instead, he proposed a "central fire" around which all celestial bodies, including the stars, Earth, Moon, Sun, and planets, revolved in circular orbits. Philolaus also hypothesized that the Earth rotates on its axis, an idea that marked a significant departure from earlier beliefs.

Eratosthenes made an extraordinary contribution by accurately measuring the Earth's radius. He achieved this by comparing the shadow cast by a pole in Alexandria with the position of the Sun at its zenith in Syene, located 5000 stadia (1 stadia equals 157.5 m) away. From his calculations, he determined the Earth's circumference to be 40,500 km—a remarkably accurate value compared to the modern measurement of 40,009 km.

Another notable figure, Aristarchus, a student of the Pythagorean tradition, made significant strides in measuring celestial distances. He estimated the Earth-Moon distance to be approximately 60 times the Earth's radius and deduced that the Moon's diameter was about one-third that of the Earth. These achievements underscored the advanced mathematical and observational skills of Greek astronomers.

Aristarchus made groundbreaking contributions by attempting to measure the distance and size of the Sun. He observed that when the Moon is half-illuminated, it forms a right triangle with the Earth and the Sun. Using basic trigonometry, he concluded that the Sun was approximately 19 times farther from the Earth than the Moon. Consequently, he estimated that the Sun's diameter was 19 times greater than that of the Moon, making its volume about 300 times larger than the Earth's.

Although Aristarchus's measurements were imprecise, leading to a significant underestimation of the Sun's size, his work was revolutionary. He demonstrated that the Sun was vastly larger than the Earth, casting doubt on the notion that the Sun orbited a smaller Earth. Based on this, Aristarchus proposed a heliocentric model in which the Earth revolved around the Sun. This arrangement also elegantly explained the seemingly erratic trajectories of Venus and Mercury.

Unfortunately, Aristarchus's heliocentric theory, which anticipated Galileo's ideas by centuries, did not gain lasting acceptance. The dominant philosophical view, influenced by the school of Aristotle, upheld the belief in Earth's absolute immobility. This geocentric model prevailed for centuries, overshadowing Aristarchus's visionary insights.

Aristotle was profoundly influenced by the stark contrast he observed between the stars, which he considered divine and unchanging, and the Earth, which he saw as imperfect and subject to change. He rejected Aristarchus's heliocentric hypothesis, relying on poorly understood astronomical arguments, and reasserted the geocentric model, placing a motionless Earth at the center of the Universe. Following his teacher Plato, Aristotle maintained that uniform circular motion was the most perfect form of movement. However, applying this idea to the observed motion of the planets posed significant challenges. To address these inconsistencies, Aristotle initially proposed a system of 26 concentric spheres centered on the Earth, which was later refined by Callippus with the addition of seven more spheres. Aristotle further expanded the model to include 55 spheres, providing a framework that appeared to resolve the complexities of celestial motion.

Apart from his flawed astronomical theories, Aristotle's most enduring legacy lies in his contributions to logic. He established the foundations of deductive reasoning through the syllogism (e.g., *1. All animals die. 2. Dogs are animals. 3. Therefore, dogs die*), the principle of identity (*A cannot be equal to non-A*), and the principle of the excluded middle (*If A is true, not-A must be false; a third option is not allowed*). These principles form the basis of mathematical and formal logic.

In the first century AD, Ptolemy built upon Aristotle's ideas, replacing the complex system of spheres with circular orbits and ensuring there were no empty spaces between them. This geocentric conception of the Universe, supported by Aristotle's authority and the phrase "ipse dixit" (*he said it*), dominated scientific thought for 15 centuries.

The Strength of Facts and the Revolution The representation of the Universe established by Aristotle and Ptolemy endured until the early 1500s but was dismantled in less than two centuries. Credit for this transformative shift goes primarily to six individuals: the Polish Nicolaus Copernicus, the

German Johannes Kepler, the Danish Tycho Brahe, the Italian Galileo Galilei, and the English scientists Isaac Newton and Edmond Halley. They challenged the dominance of contemporary academics who clung to Aristotelian and Ptolemaic theories, as well as the broader intellectual climate, which regarded any contradiction of Aristotle's physics as scandalous. These pioneers also had to overcome their own biases, breaking free from entrenched ideas of perfection and beauty, such as the belief in uniform circular motion, spherical shapes, the immobility of stars, the finiteness of the cosmos, the existence of an Empyrean heaven hosting God's physical presence, and the literal interpretation of the Old Testament. Additional challenges included grappling with the concept of forces acting over a distance—an idea that was difficult for many to accept. The true turning point came with a shift in approach: from conceptualizing nature through abstract reasoning alone to actively questioning it through observation and experimentation. Galileo Galilei articulated this shift most explicitly and publicly. His method, now known as the scientific method, established that understanding nature requires questioning it and analyzing its responses. This marked the definitive introduction of the concept of scientific experimentation. In astronomy, this meant systematic observation, while in other sciences, it involved designing experiments to investigate specific phenomena. This approach laid the foundation for the rapid development of science that continues to this day.

Nicolaus Copernicus was born into a wealthy family. Orphaned at a young age, he was raised by his uncle, a bishop, who sent him to study at the University of Bologna. There, alongside studying canon law, Copernicus attended astronomy lectures given by the prominent astronomer Domenico Maria Novara. Upon returning to his homeland, he delved deeply into Ptolemy's writings but was troubled by the convoluted and unwieldy solutions devised to explain planetary motions in the solar system. Through mathematical calculations, he realized that a heliocentric model could significantly simplify the interpretation of astronomical observations. However, Copernicus remained committed to the ancient idea of uniform circular motion as the most perfect form of movement. To reconcile this belief with his heliocentric model, he introduced a number of complex adjustments. The heliocentric hypothesis, which placed the Earth in orbit around the Sun, also required the acceptance of an enormously larger cosmos to account for the apparent immobility of the stars. This posed a direct challenge to the concept of the Empyrean, further complicating the acceptance of his ideas.

Because of these theological and philosophical implications, Copernicus delayed publishing his seminal work, *De Revolutionibus Orbium Coelestium* (*On the Revolutions of the Celestial Spheres*—Fig. 2.6), until he was on his

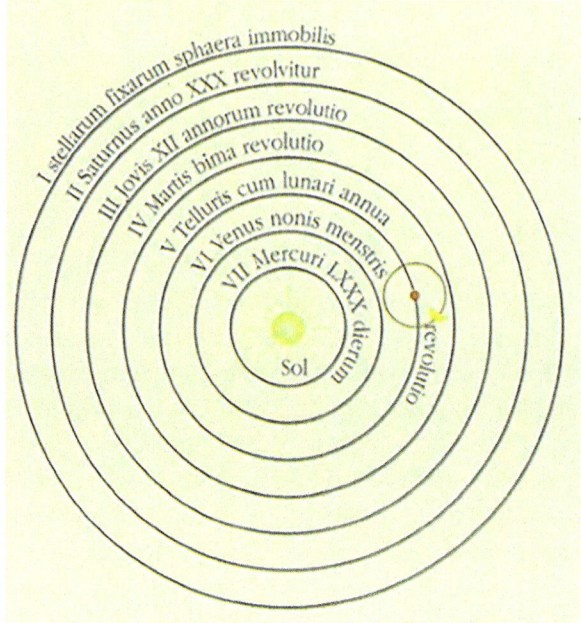

Fig. 2.6 Copernican diagram included in Copernicus's book *De revolutionibus orbium coelestium*; at that time, scientists from every nation wrote in Latin

deathbed. Despite these challenges, his ideas set the stage for a scientific revolution that would transform humanity's understanding of the cosmos.

Some years later, a young Danish astrologer, Tycho Brahe, made significant observations while studying the sky. He discovered a remarkably bright new star, unlike anything previously recorded, as well as a comet that he estimated to be very distant—beyond the orbit of Venus. He noted that the comet's tail always pointed in the opposite direction from the Sun. These observations shattered the long-held belief that the heavens were immutable and unchanging. Like many scientists of his era, Tycho was not financially constrained. His family was so wealthy that King Frederick II, in an effort to settle a substantial debt owed to the Brahe family, granted Tycho an entire island, complete with a castle, servants, and state-of-the-art astronomical instruments.

Although Tycho was intrigued by the Copernican model, he lacked the conviction to fully embrace the idea of a moving Earth. Instead, he proposed a hybrid system that maintained Earth's centrality and immobility, with the Sun and Moon orbiting the Earth, while the other planets orbited the Sun (Fig. 2.7). This geocentric compromise, however, introduced challenges, such as explaining how celestial bodies could traverse the supposedly non-transparent

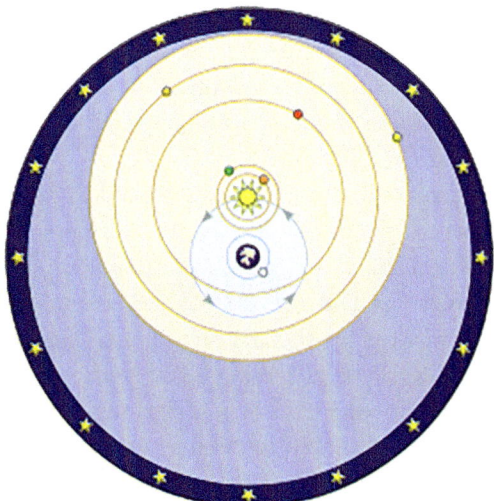

Fig. 2.7 Diagram of Tycho Brahe's System. In Tycho's model, the Earth is once again placed at the center of the cosmos, with the Moon and the Sun revolving around it in circular orbits. Meanwhile, Mercury, Venus, Mars, Jupiter, and Saturn orbit the Sun, which in turn orbits the Earth, along with the fixed stars. This model represents an attempt to bridge traditional geocentric views with the emerging heliocentric ideas of the time. (*Source: Fastission-Wikipedia*)

interplanetary spaces. Despite these limitations, Tycho's precise and detailed observations laid the groundwork for future breakthroughs in astronomy.

Around the beginning of the 1600s, Tycho Brahe invited a young German astronomer, already recognized for his discoveries despite his poverty and frail health, to work as his assistant. This young man, Johannes Kepler, had studied at the seminary of the University of Tübingen, where, in addition to theology, he was introduced to mathematics and the Copernican model of the cosmos. Kepler found the Copernican system far more convincing than the Ptolemaic model. He also came to understand that the Sun played a crucial role in driving the motion of the planets and that an unknown force caused the planets to move faster when closer to the Sun.

Before his death, Tycho Brahe entrusted Kepler with his extensive and meticulously collected astronomical data. Building on the idea of a moving Earth, Kepler attempted to refine the explanation of Mars's orbit, although his initial calculations could not yet fully account for its motion. However, through persistent observation and analysis, Kepler made a groundbreaking realization: planetary orbits are not circular but elliptical, with the Sun located at one of the foci.

This revolutionary insight resolved many inconsistencies and effectively dismantled the Ptolemaic system, paving the way for a new understanding of celestial mechanics (in Fig. 2.8 the monument to Tycho Brahe and Johannes Kepler, located in Prague, is depicted, along with the frontispiece of one of Kepler's most important publications).

A few years after Kepler's discoveries, Galileo Galilei was conducting his own groundbreaking observations of the sky. Using a telescope with a modest magnification of just 20×, equipped with new lenses crafted in Murano near Venice. He pointed the telescope in the direction of the Moon and observed that its surface was irregular, marked by undulations, circular cavities, and mountains. These features contradicted the prevailing belief in the celestial perfection and immutability of heavenly bodies. As a result, the long-held distinction between the earthly realm and the celestial quintessence was irrevocably challenged, dismantling yet another assumption that had persisted for centuries.

At the time, Galileo was a professor at the University of Padua and spent his nights observing the heavens. After studying the Moon, he directed his

Fig. 2.8 On the left is the monument dedicated to Kepler and Brahe, located on a hill in Prague (*Source: Wikipedia Creative Commons Attribution-Share Alike 3.0 Unported license*). On the right is one of Kepler's most significant publications

telescope toward Jupiter and discovered small points of light surrounding the planet. By tracking their movements, he realized these were moons orbiting Jupiter. This discovery refuted the argument that if Earth were in motion, it would lose its own Moon. Additionally, Galileo noted that Jupiter and other planets appeared as small disks through the telescope, while even the brightest stars remained point-like. This distinction convinced him unequivocally of the heliocentric model and the existence of an immense void between Saturn and the stars (in Fig. 2.9 the frontispiece of what is perhaps Galileo's most important publication and his telescopes).

Galileo faced significant opposition to his findings. The scientific, religious, and philosophical worldview of the era was deeply entrenched, and his ideas clashed with the Church's teachings. The Church rejected the concept of a

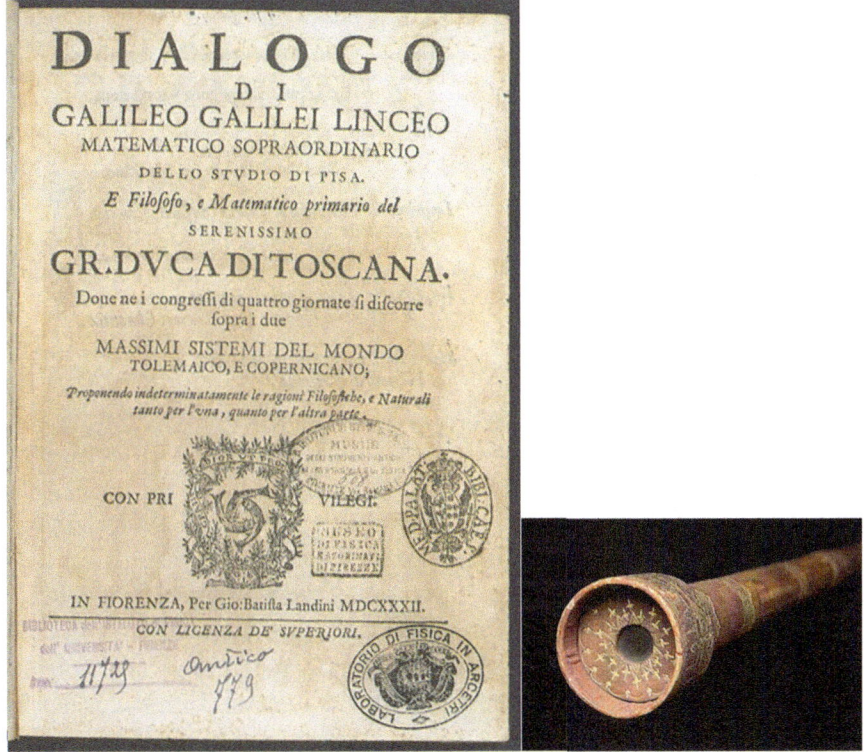

Fig. 2.9 On the left is the frontispiece of *"Dialogue Concerning the Two Chief World Systems"-1632 (Dialogo sopra i due massimi sistemi del mondo)* where Galileo systematically disproves the Ptolemaic model; at the beginning of his scientific career, Galileo wrote in Latin, but later he adopted Italian. On the right is one of the telescopes used by Galileo. (*Source: Wikipedia- licensed under the Creative Commons Attribution-Share Alike 4.0 International license*)

moving Earth, as it contradicted a literal interpretation of the Old Testament. However, Galileo stood firm, asserting that facts could not be altered by authority. Ultimately, 180 years after Galileo's death, the Church declared the heliocentric model fully compatible with Christian faith.

As a devout Christian, Galileo sought to reconcile his discoveries with his beliefs. In a letter to Grand Duchess Christina of Lorraine, he wrote:

> Here I would say what I understood from an ecclesiastical person of very eminent degree: the intention of the Holy Spirit is to teach us how to go to heaven, not how the heaven goes. (Io qui direi quello che intesi da persona ecclesiastica costituito in eminentissimo grado cioè l'intenzione dello Spirito Santo essere di insegnarci come si vadia al cielo, e non come vadia il Cielo.)

Universal Gravitation: The Key to Understanding the Universe The new representation of the universe was gradually taking shape, but a unifying explanation was still needed—enter Isaac Newton. Newton attended Trinity College to study theology but was far more captivated by the works of Copernicus, Galileo, Kepler, and Descartes than by the Aristotelian physics prescribed in his curriculum. In 1665, when a plague epidemic struck London, Newton returned to his hometown in Lincolnshire, where his enforced isolation became a period of extraordinary creativity. During this time, Newton developed differential calculus to analyze and understand the solar system. Although calculus is now formulated differently, Newton's approach remained a cornerstone of astronomy for centuries.

His most profound discovery, however, was the law of universal gravitation. Building on Galileo's earlier experiments with falling objects, Newton established that gravitation was a force attracting bodies toward each other. Galileo had already demonstrated that this force caused acceleration, not constant velocity as Aristotle had claimed. Newton expanded this understanding, showing that the gravitational force increases with mass and proximity. He also incorporated Galileo's principle of inertia: a body in motion, unaffected by external forces, continues in its trajectory. Newton applied these principles to the motions of the Sun and planets, abandoning the Aristotelian-Ptolemaic notion of uniform circular planetary motion in favor of elliptical orbits. He further established that the force acting on a body is proportional to the product of its mass and acceleration. However, one unresolved issue remained: how could forces act over a distance? Descartes had proposed the existence of an all-pervading "ether" to address this, but Newton, avoiding speculation, attributed this mystery to divine intervention, writing:

> The motions which the Planets now have could not spring from any natural cause alone but were impressed by an intelligent Agent.

Newton also humbly acknowledged the giants on whose work he built, famously stating:

> If I have seen further, it is because I was standing on the shoulders of Giants.

Newton's laws were powerfully validated by observations of Halley's Comet. Using these laws, Halley calculated the comet's motion, proposing it followed an elliptical or parabolic orbit. Comparing historical records, he deduced that sightings in 1531, 1607, and 1682 were of the same comet, appearing approximately every 75 years. Predicting its return on April 13, 1759, Halley's calculations proved accurate. Historical evidence traced the comet back to 1301 (observed by Giotto see Fig. 2.10 left), 837, 12 BCE (potentially the "Star of Bethlehem"), and 467 BCE, offering compelling confirmation of Newton's universal laws (in Fig. 2.10 right a picture of the Halley comet appeared in 1911).

Halley also used Newtonian mechanics to predict Venus's transit across the Sun. Although he did not live to see it, French and English astronomers

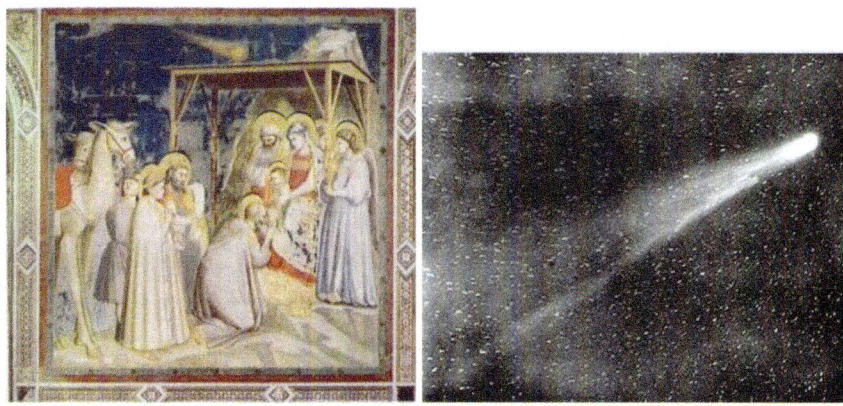

Fig. 2.10 *Left:* Giotto's *Adoration of the Magi* (1305) in the Scrovegni Chapel, Padua, inspired by his observation of Halley's Comet in 1301, which he depicted as the Star of Bethlehem. *Right:* A photograph of Halley's Comet during its 1911 appearance. (*Source: Trentino Cultura accadeoggi*)

observed the event on June 6, 1761, validating his calculations. These observations further enabled the measurement of the Earth-Sun distance, estimated at 153 million kilometers—a remarkably accurate figure compared to modern measurements.

An Infinite Number of Stars: Unraveling the Mysteries of the Universe A significant challenge in the Newtonian Universe was the simple yet profound question: *Why is the night sky dark?* This paradox hinted at the vastness of the universe. As Johannes Kepler pointed out, if the cosmos was infinitely filled with stars, the night sky would glow brilliantly. Following the establishment of Newton's universal gravitation, which explained planetary and cometary motions, astronomers began investigating realms beyond the solar system. In 1782, William Herschel, alongside his sister Caroline, constructed a revolutionary telescope capable of gathering 1000 times more light than Galileo's. Over 20 years, they meticulously cataloged hundreds of thousands of stars, producing the first map of the Milky Way. Their work suggested the galaxy contained millions of stars.

While studying double stars—stellar pairs appearing close in the sky—the Herschel observed changes in the angle of the line joining the stars, indicating mutual orbits. Confirmation that these orbits adhered to Newton's laws came later through William's son, John Herschel, further affirming the universal applicability of Newtonian gravitation.

The Herschel also identified objects distinct from stars and planets, which they termed *nebulae*. Through extensive observations, they hypothesized these nebulae were diffuse gas clouds condensing into star clusters under the influence of gravity.

Technological Advances: Photography and Spectroscopy The nineteenth century brought two groundbreaking tools: photography and spectroscopy. Photography enabled astronomers to collect light over extended periods, capturing faint celestial objects on photographic plates with exceptional clarity. Spectroscopy, on the other hand, allowed scientists to dissect light into its component wavelengths using optical prisms. Light is an electromagnetic wave and varies in color based on its frequency—the number of oscillations per second. When sunlight is passed through a prism, it disperses into a chromatic sequence from red to violet, creating what Huygens termed the *spectrum*.

In 1814, German physicist Joseph Fraunhofer studied the Sun's spectrum and identified dark lines, which he deduced were caused by gases in the solar atmosphere. This technique, extended to nebulae, confirmed they consisted of gas. Spectroscopic analysis also revealed the chemical compositions of stars, showing they contained elements identical to those found on Earth (in Fig. 2.11 an example of nebula).

Spectroscopy and Atomic Insights By the mid-1800s, scientists recognized that each chemical element produced a unique spectral pattern, akin to a fingerprint. Although the precise connection between spectra and atomic composition was unclear, the groundwork was laid. In the early twentieth century, Danish physicist Niels Bohr, building on experiments by Sir Joseph John Thomson, James Chadwick, and others, unraveled the relationship between atomic structure and spectral characteristics, completing this crucial puzzle. Another critical breakthrough came from the analysis of starlight to determine the chemical elements present in stars. This process, known as spectroscopic analysis, involves examining the characteristics of the light waves emitted by stars. By analyzing

Fig. 2.11 The Thor's Helmet Nebula featuring swirling clouds of green and red gas with numerous bright stars scattered throughout. The nebula's intricate patterns and colors create a dynamic and ethereal scene against the dark backdrop of the universe. (*Source: ESO*)

these waves, scientists can identify the elements in a star, which are remarkably similar to those found on Earth (in Fig. 2.12 an example of spectrum).

Gradually, pieces of knowledge about the cosmos began to fall into place. However, the understanding of the vast distances between stars remained incomplete, and there was still no definitive proof that stars existed beyond the Milky Way. The breakthrough came from an American astronomer, Henrietta Leavitt, who worked at the Harvard Observatory. Leavitt's task was to analyze thousands of photographic plates in search of stars whose brightness varied periodically over time. Specifically, she focused on a class of variable stars known as Cepheids in the Small Magellanic Cloud, where she discovered a crucial relationship: the period of variability in a Cepheid's brightness was directly linked to its intrinsic luminosity.

This discovery was pivotal. Between 1919 and 1924, Edwin Hubble used the powerful Mount Wilson telescope to observe Cepheids in distant galaxies. By measuring the distance to one of these stars, Hubble calculated it was approximately 930,000 light years away. A second Cepheid gave a similar result of 850,000 light years. This proved that these stars were not part of the Milky Way, but were instead located in galaxies beyond our own.

This discovery of the vast scale of the Universe helped explain a long-standing puzzle: why is the night sky dark? The answer is that these stars are incredibly far from us and from each other, making their light insufficient to fill the night with brightness. The realization of the cosmos' enormous dimensions was a key milestone in our understanding of the Universe.

The Cosmos The systematic study of the stars has revealed that the universe contains billions of galaxies, each with billions of stars. However, our quest to understand the Universe is far from over. A crucial breakthrough in this understanding came from an effect named after the Austrian physicist Christian Doppler. This effect refers to the change in the color of light emitted by stars depending on their movement relative to us: if a star is moving toward

Fig. 2.12 A spectrum image (iron) displaying vertical lines of varying colors on a black background. The lines transition from blue on the left, through green and yellow in the center, to red on the right. The colors represent different wavelengths of light, illustrating a visible light spectrum

us, the light shifts toward the blue end of the spectrum; if it is moving away, the light shifts toward the red end. For decades, scientists have known that the universe is expanding, and this discovery was further supported by the Doppler effect. The shift of light toward the red end of the spectrum (redshift) indicates that stars and all celestial bodies are moving away from one another. More recently, through satellite observations equipped with advanced detectors, we have learned that not only is the universe expanding, but this expansion is also accelerating.

I'll conclude this very brief and incomplete overview of humanity's evolving understanding of the sky here. Beyond these discoveries lie concepts like general relativity, the Big Bang model, the accelerated expansion of the universe, dark matter, and dark energy.

2.3 A Missing Piece

In the relentless pursuit of understanding the Universe and its composition, many questions remain unanswered. As some of you may already know, recent discoveries have suggested the existence of dark matter and dark energy. These components are thought to dominate the cosmos, reducing the fraction of the Universe we have traditionally studied and observed to just 5%.

Among the unresolved mysteries, one pivotal question has persisted: what mechanisms generate the energy that powers the Sun and stars, allowing them to shine so brilliantly?

In the 1930s, a German physicist Hans Bethe proposed hypotheses that addressed this question. He suggested that the energy in the Sun is produced through a series of nuclear fusion reactions, beginning with the fusion of hydrogen nuclei and progressing to other nuclear reactions.

Nuclear fusion involves the combination of two light nuclei to form a heavier one. The resulting nucleus has a mass slightly less than the sum of the masses of the original nuclei. For example, when four hydrogen nuclei fuse to form a helium nucleus, the helium's mass is less than the combined masses of the four hydrogen nuclei. This "missing" mass is converted into energy, which is released.

This energy is measured in specialized units called electron-volts (eV), which represent the kinetic energy gained by an electron moving through an electric field with a potential difference of 1 V. Since an eV corresponds to a very small amount of energy—1.6×10^{-19} J—larger multiples such as kilo-electron-volts (keV), megaelectron-volts (MeV), and gigaelectron-volts (GeV) are often used in physics. The joule is the unit of measurement for energy in the International System of Units (SI). 10^{-19} means one tenth of a billionth of a billionth.

An atom, a fundamental building block of matter, consists of a positively charged nucleus surrounded by negatively charged electrons. The nucleus, much smaller than the atom itself, contains protons (positively charged particles) and neutrons (neutral particles), collectively called nucleons. The number of protons in a nucleus determines the element, while variations in the number of neutrons result in isotopes, which can sometimes be radioactive.

Bethe's nuclear fusion hypothesis for the Sun begins with the fusion of two hydrogen nuclei to form deuterium, an isotope of hydrogen comprising one proton and one neutron. This process, called *pp reaction* is the starting point for further fusion reactions (*pp chain*). Among these are reactions involving the fusion of hydrogen nuclei with electrons (pep reactions), a reaction between a beryllium nucleus (Be-7) and an electron, and a reaction involving a boron nucleus (B-8). These four reactions emit detectable neutrinos, particles that provide critical evidence of fusion processes through their measurable energies.

The pp chain reaction generates a core temperature of approximately 15 million degrees Celsius—sufficient to maintain the Sun's structure. This extreme heat creates pressure that counteracts the Sun's immense gravitational force, preventing it from collapsing under its own weight. However, in massive stars—those with a mass at least 30% greater than the Sun—the gravitational force is significantly stronger. The pp chain alone cannot sustain these stars, as it does in the Sun. Bethe and Carl von Weizsäcker hypothesized in the 1939 another mechanism: the *CNO cycle*. This cycle operates at temperatures ten times greater than the Sun's core and involves hydrogen fusion facilitated by carbon, nitrogen, and oxygen as catalysts. Catalysts are substances that accelerate reactions without being consumed, and in this case, they significantly increase the rate of fusion. In this cycle reactions that emit significant fluxes of neutrinos are two, one involving nitrogen N-13 and the other oxygen O-15.

In the Annexes 2.1 and 2.2, the reactions occurring in the chain pp and the cycle CNO, respectively, are detailed.

2.4 Neutrinos and Solar Neutrinos, Formidable Probes

Let us begin with this question: how can we study the Sun and the stars when, due to their extremely high temperatures, no human instrument can penetrate them for direct study? Probes capable of resisting such intense heat and conditions without melting have never been invented. So, how have we managed to gather insights about these celestial bodies? What ingenious solution have we devised to tackle this challenge? The truth is, we didn't invent anything new—instead, we leveraged the unique properties of a fundamental elementary particle known as the *neutrino*.

But first, what are elementary particles? They are the fundamental building blocks of matter, the most basic components beyond which nothing smaller exists. Only 12 particles, along with their corresponding antiparticles, are truly elementary. These particles, together with three of the four fundamental forces of nature—namely the strong nuclear force, the weak nuclear force, and the electromagnetic force (the gravitational force is too weak to play a significant role in the fundamental structure of matter)—form the basic framework of matter. They govern the physical laws that make the matter we know operate. Among these 12 particles, the neutrino stands out as particularly unique.

Neutrinos are elementary particles so incredibly small that we have yet to measure their exact size. However, we know their diameter is less than a billionth of a billionth of a meter. They have no electric charge and an extraordinarily low probability of interacting with matter. Neutrinos can pass through the entire Sun, the stars, and even the vast expanse of the Universe without being significantly affected. These properties—especially their ability to traverse inaccessible environments—make neutrinos extraordinary probes. By reaching us from places like the Sun's core, distant stars, or even the Earth's interior, neutrinos carry invaluable information about the processes occurring in these otherwise unreachable regions.

Neutrinos come in three forms, often referred to as *"flavors,"* which are similar and distinct (Fig. 2.13). They are connected to three other elementary particles belonging to the *lepton family* (from the Greek word *leptós*, meaning "light"), a group that also includes the electron. The leptonic particles consist of the electron, the muon (μ, pronounced *"mu"* from the Greek alphabet and is called also *muon*), the tau (τ), and their corresponding neutrinos: the electron-neutrino, muon-neutrino, and tau-neutrino. According to all current theories and evidence, these neutrino flavors were once believed to be

Fig. 2.13 A humorous depiction of the neutrino family: electron-neutrino, muon-neutrino, and tau-neutrino. The muon (represented by the Greek letter μ) and the tau (represented by the Greek letter τ), like the electron, are elementary particles belonging to the lepton family—one of the two major families of elementary particles

immutable—meaning that an electron-neutrino, for example, would always remain an electron-neutrino, and the same held true for the others. Their distinct identities, or flavors, are a cornerstone of our understanding of these fascinating particles.

The belief that neutrino flavors could not change was a core tenet of what is known as the Standard Model of particle physics. This model describes and predicts the properties of elementary particles—the fundamental building blocks of matter that cannot be broken down into smaller components. The Standard Model is constantly refined as new particle properties and phenomena are discovered.

Such was the case with neutrinos, following the discovery of an intriguing phenomenon: during their 8-min journey from the Sun to Earth, electron-neutrinos (produced in the Sun's nuclear reactions) can transform into muon-neutrinos (μ-neutrinos) or tau-neutrinos (τ-neutrinos). This phenomenon, known as *neutrino oscillation*, had profound implications. Before this discovery (observed for the first time in atmospheric neutrinos by the Super-Kamiokande experiment), the Standard Model assumed neutrinos were massless. However, neutrino oscillation can only occur if neutrinos possess a nonzero mass. Although we now know neutrinos have mass, their mass is so minuscule that it has not yet been directly measured.

The Sun bombards the Earth with an enormous number of neutrinos (Figs. 2.14 and 2.15): every second, approximately 60 billion neutrinos pass through a single square centimeter of your thumbnail!

When I worked at the Gran Sasso Laboratory, I occasionally participated in outreach events to explain our research to non-scientific audiences. During one such event, while discussing solar neutrinos, a concerned attendee asked whether these neutrinos might be responsible for increasing infertility rates. I reassured him that the extremely low probability of neutrinos interacting with matter means they have no harmful effects on the human body. In fact, solar neutrinos have been streaming through humanity since the dawn of our species without causing any damage.

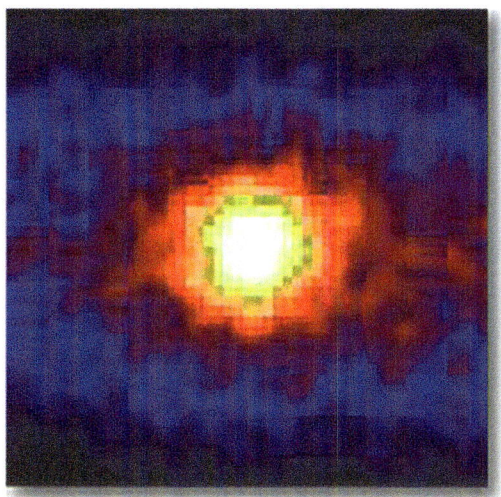

Fig. 2.14 The Sun "seen" through the neutrinos it emits. This is not a photograph but an electronic recording. The neutrinos are detected using the Japanese Super-Kamiokande detector, which contains approximately 45 million liters of water. When neutrinos—extremely rarely—interact with the water, they collide with an electron, transferring energy to it. This interaction produces light through a phenomenon called Cherenkov radiation, named after its discoverer. The emitted light is converted into an electrical pulse, electronically amplified, and then processed by a computer to create the image. Remarkably, this neutrino-based image provides a direct glimpse into the Sun's core, or nucleus (yellow and green) and reveals the outermost layer of the Sun—the solar photosphere, or surface (red). This image appears inherently blurred because each point is reconstructed with at least a minimal degree of uncertainty, as is the case with anything reconstructed experimentally. (*Source: Super-Kamiokande*)

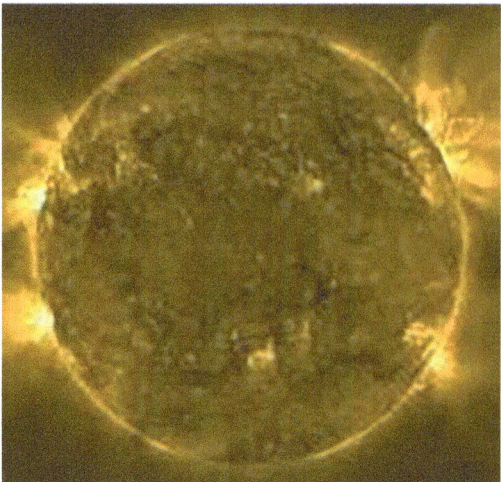

Fig. 2.15 The Sun as seen through the light it emits, showcasing its surface with visible solar flares and prominences. The image appears blurred not due to low resolution but because it is inherently characteristic of this type of representation. (*Source: CREDIT ESA & NASA/Solar Orbiter/EUI team; Data processing: E. Kraaikamp (ROB)*)

Studying solar neutrinos, which are produced in the Sun by nuclear fusion reactions and subsequently emitted, is no easy task. While the number of neutrinos is extraordinarily high, their interactions with matter are exceedingly rare, making it difficult to detect signals caused by them. Because neutrinos are electrically neutral, they do not produce direct signals. To observe them, they must collide with a charged particle—typically electrons present in the matter—which then travels a short distance, releasing the energy it acquired from the interaction.

Annex 2.1: Nuclear Fusion Reactions in the pp Chain operating in the Sun

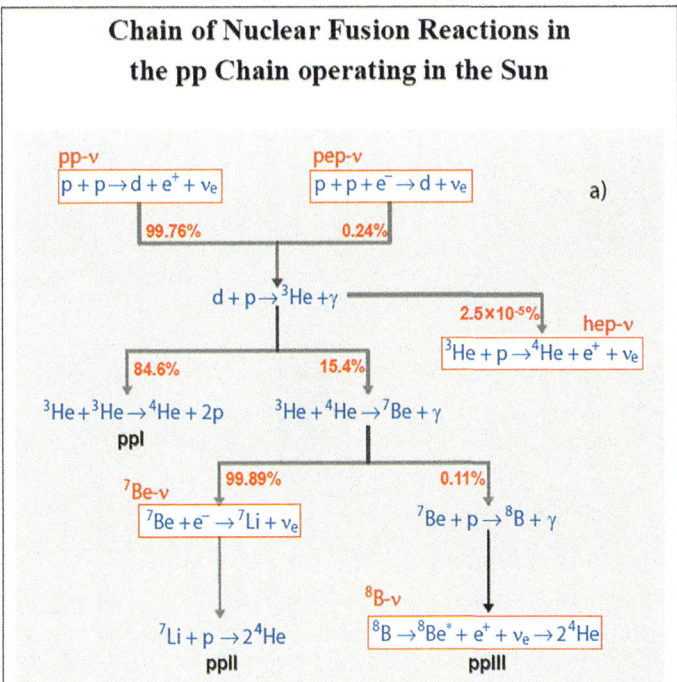

The pp chain accounts for 99% of the Sun's energy production. This series of nuclear reactions begins the fusion of two hydrogen nuclei and continues through interactions involving increasingly heavier nuclei. In the accompanying figure, five key reactions are highlighted in white. Each of these reactions emits neutrinos represented by the Greek letter ν, with the subscript e, known as the electron-neutrino (see text).

Diagram titled "Chain of Nuclear Fusion Reactions in the pp Chain operating in the Sun," illustrating the sequence of reactions in the proton-proton chain. The process begins with two pathways: pp-ν and pep-ν, leading to the formation of deuterium. Subsequent reactions involve helium-3 and helium-4, with branching paths showing probabilities of 99.76%, 0.24%, 84.6%, 15.4%, and 0.11%. Key reactions include the formation of beryllium-7, lithium-7, and boron-8, with neutrino emissions indicated by the Greek letter ν. The diagram highlights the main reactions contributing to the Sun's energy production.

Annex 2.2: The CNO Cycle

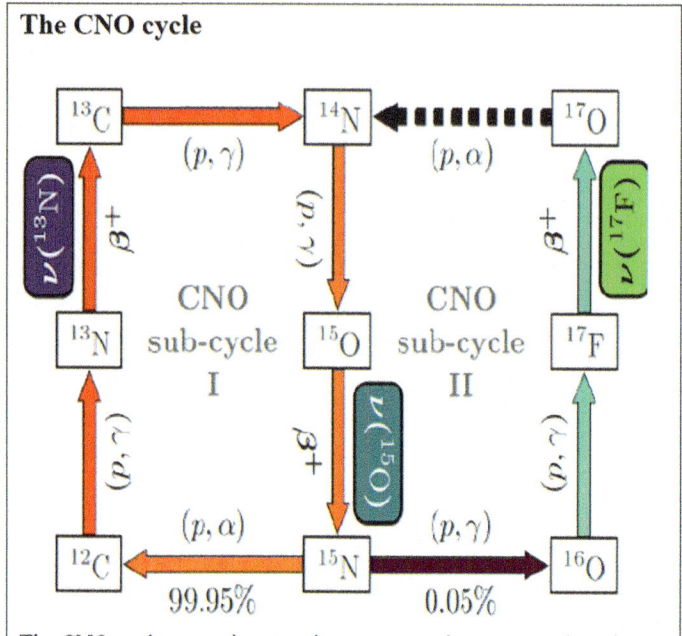

The CNO cycle comprises two interconnected processes: the primary cycle, depicted on the left, which overwhelmingly dominates at 99.95%. Within this primary cycle, reactions involving nitrogen and oxygen result in the production of electron-neutrinos. The second cycle contributes only at 0.05%.

Diagram illustrating the CNO cycle, a process in stellar nucleosynthesis. It consists of two sub-cycles. The primary cycle, on the left, involves carbon, nitrogen, and oxygen transformations, dominating at 99.95%. Key reactions include proton capture and beta decay, producing electron-neutrinos. The secondary cycle, on the right, contributes 0.05% and involves additional transformations between nitrogen, oxygen, and fluorine. Arrows indicate reaction pathways, with labels such as (p, γ) and (β+).

3

The Genesis of the Experiment

On a foggy February evening, late in the afternoon, an Indian-American physicist named Raju Raghavan, from the prestigious Bell Labs in New Jersey, knocked on the door of my office at the Department of Physics, University of Milan. He had come to propose an experiment on solar neutrinos—neutrinos emitted by the Sun. Accompanying him was another American physicist whose name I cannot recall and whom I never met again.

Raju had initially come to Milan to speak with my colleague Ettore Fiorini, a physicist who had studied neutrinos since the beginning of his career. At the time, however, Fiorini was not interested in a solar neutrino experiment. He referred Raju to me, as I had experience coordinating large collaborations and was, at that time, the Principal Investigator (PI) of an Italian collaboration involving three institutes for an experiment at Fermilab (Fermi National Accelerator Laboratory) near Chicago.

I wondered why Raju trusted me with such a proposal. He explained that he had received very positive feedback about my Milan team from his contacts at Fermilab. His proposal involved a large-scale experiment requiring about 10,000 tons of scintillator—a material that emits light when a particle loses energy within it. However, the experiment would only detect a small fraction of the solar neutrino flux, about one-thousandth of the total. This limitation stemmed from the presence of natural radioactivity in all materials, which, if not sufficiently suppressed, would obscure the rare neutrino events. As previously mentioned, neutrinos interact very weakly with matter, making their detection extremely challenging.

I had never worked with neutrinos, let alone solar neutrinos, which involved processes at very low energies. In contrast, my work with particle accelerators dealt with much higher energies, typically in the range of hundreds of GeV. However, I found the proposal intriguing, particularly because of the heightened interest at the time in solar neutrinos and the so-called *Solar Neutrino Problem*, a subject of intense scientific debate.

3.1 The Solar Neutrino Problem

The Solar Neutrino Problem arose from an incorrect interpretation of results obtained by four different experiments conducted in Japan, Italy, Russia, and the United States. The misunderstanding was due to the lack of knowledge about a property of neutrinos that had not yet been discovered. The issue was brought to prominence by the experiment designed and built by Raymond Davis Jr. in the Homestake gold mine in Lead, South Dakota, which operated from 1970 to 1994.

You might wonder why an experiment studying neutrinos from the Sun would be conducted underground in a mine. It may seem counterintuitive to study the Sun by going underground. A humanist friend of mine once joked that physicists were a bit crazy, always chasing fanciful ideas and spending large sums of money on them.

In reality, there is a very logical reason for conducting such experiments underground. Detecting the rare interactions of neutrinos requires shielding from the vast number of cosmic rays that constantly bombard the Earth's surface (In Fig. 3.1 a representation of cosmic rays). These cosmic rays would overwhelm detectors placed on the surface, masking the few signals from neutrino interactions. By placing experiments in underground laboratories, the surrounding rock serves as a natural shield, absorbing nearly all cosmic rays except neutrinos and, to a lesser extent, muons (μ particles), which are also partially absorbed.

While an experiment sought to measure the flux of solar neutrinos on Earth, American physicist John N. Bahcall developed a model that became the cornerstone of all solar research: the "Standard Solar Model" (SSM). This model was refined over 15 years and continued to evolve with contributions from various physicists under Bahcall's guidance. The SSM assumes that the Sun is spherically symmetric, with a slowly rotating core, uniform magnetic fields, and an initial composition of approximately 71% hydrogen, 27% helium, and 2% heavier elements—values corresponding to the current chemical composition of the solar photosphere. The model is designed to

Fig. 3.1 Cosmic rays reaching Earth are particles generated in the upper atmosphere when high-energy particles from space collide with atoms and molecules in the atmosphere. These collisions produce secondary particles and radiation that eventually reach the Earth's surface. (*Source: ASPERA/Novapix/L. Bret-Spazi Culturali—INFN*)

satisfy observable solar properties, providing accurate predictions for the Sun's mass, luminosity, and surface temperature.

The Solar Neutrino Problem arose from Ray Davis Jr.'s experiment, which sought to measure the solar neutrino flux. The results revealed only one-third of the neutrino flux predicted by the Standard Solar Model (SSM), creating a discrepancy that could not be explained without assuming significant errors in either the experimental setup or the theoretical model.

Over a decade later, the Japanese Kamiokande experiment, located in the Kamioka mine, reported similar findings. In addition, Kamiokande had the advantage of confirming the solar origin of the detected neutrinos. This was possible because the experiment utilized Cherenkov radiation—light produced by relativistic particles (particles traveling at a significant fraction of the speed of light), such as electrons struck by neutrinos. The directionality of Cherenkov light allowed researchers to trace the electrons' paths, which partially retained the direction of the incoming neutrinos.

This measurement was later revisited by the GALLEX and SAGE experiments an Italian-German collaboration at the Gran Sasso Underground Laboratory in Italy and a Russian-American experiment at Baksan Neutrino Observatory in Russia. They measured about 60% of the solar neutrino flux expected from the SSM.

The resolution of the Solar Neutrino Problem was directly tied to the phenomenon of neutrino oscillation already discovered by Super-Kamiokande in 1998 studying the atmospheric neutrinos. The decisive experiment was the

Sudbury Neutrino Observatory (SNO) in Canada, which operated from 1999 to 2013 (Fig. 3.2). SNO used a detector containing 1000 tons of heavy water and exploited Cherenkov light produced by relativistic particles. Heavy water is a compound similar to regular water where the hydrogen, which consists of a positively charged proton, is replaced by deuterium. Deuterium is an isotope of hydrogen where the nucleus contains one *proton* and one electrically neutral *neutron;* therefore, while water is composed of two hydrogen nuclei and one oxygen nucleus (H_2O), heavy water is composed of two deuterium nuclei and one oxygen nucleus (D_2O). The experiment was installed 2 km underground in the Creighton Mine near Sudbury, Ontario.

These Cherenkov detectors were able to measure solar neutrinos with energies of 5 MeV and above. This limitation arises from the natural radioactivity present in all materials. Although not harmful to human health, this background radiation is sufficient to obscure neutrino events. Consequently, it is not enough to place detectors underground to shield them from cosmic rays; emissions from radioactive nuclei in the materials and in the rocks, must also

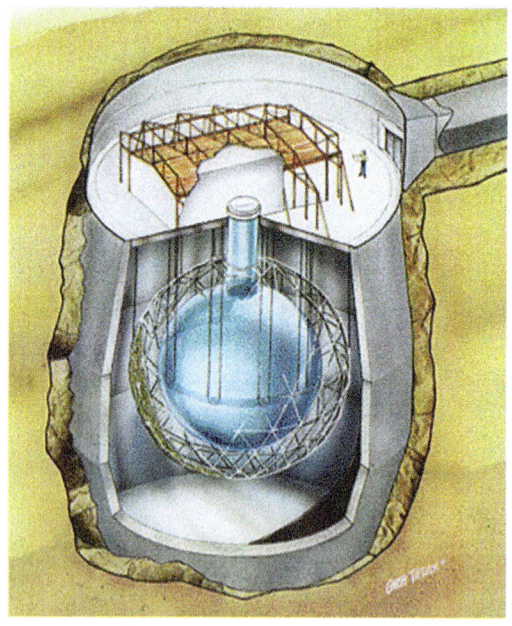

Fig. 3.2 SNO Experiment: the figure shows the central container housing the heavy water used to detect neutrino interactions, the surrounding structure supporting the "electronic eyes" (photomultipliers) that capture the light produced within the central container, and the large external mechanical framework filled with water to shield against external radiation. (*Source: SNO Site*)

be minimized. The decay products of natural radioactive elements generally have energies below 5 MeV, restricting neutrino studies to energies above this threshold. To detect neutrinos with lower energies, natural radioactivity must be reduced to nearly zero. Additionally, emissions from the surrounding rocks and underground atmosphere must be addressed. Radon-222, a radioactive gas emitted by rocks, is a particular challenge and must be effectively shielded to prevent interference with detector measurements.

The Davis, GALLEX and SAGE experiments, which observed a solar neutrino flux significantly lower than predicted by the Standard Solar Model (SSM), relied on the transitions of specific nuclei in their detectors. Crucially, these transitions could only be triggered by one specific type of neutrino: the electron neutrino. Solar neutrinos are initially produced as electron neutrinos, but during their journey from the Sun to Earth, some transform into muon-neutrinos or tau-neutrinos.

The Sudbury Neutrino Observatory (SNO) resolved the solar neutrino problem because it could detect all three types of neutrinos—electron, muon, and tau. By observing the total flux of all three flavors neutrinos, the experiment demonstrated that the combined interactions matched the predictions of John Bahcall's Standard Solar Model.

This groundbreaking result from the Canadian SNO experiment not only solved in 2002 the solar neutrino problem but confirmed in solar neutrinos the phenomenon of neutrino oscillation (after its discovery in atmospheric neutrinos by Super-Kamiokande). The phenomenon of neutrino oscillation violates a key rule predicted by the Standard Model of particle physics: that neutrinos cannot transform into different types (or "flavors"). The phenomenon of neutrino oscillation, where neutrinos change flavors during their journey, had been theoretically proposed earlier by the physicist Bruno Pontecorvo.

More or less simultaneously with SNO, the Super-Kamiokande experiment was operating in Japan, studying solar neutrinos through Cherenkov light, not in heavy water like SNO, but in ordinary water. Super-Kamiokande faced more challenges than SNO in measuring the flux of neutrinos of all types because it encountered greater difficulties in detecting neutrinos of all flavors.

3.2 Borex and Borexino

The experiment Raghavan proposed was a large-scale based on scintillator liquid containing Boron. He suggested naming it **Borex** (short for Boron Experiment). Due to natural radioactivity, this experiment could only study solar neutrinos with energies above 5 MeV, similar to other experiments that were under construction during those years, such as SNO and Super-Kamiokande, even though we would have had some advantages, such as better resolution because the scintillator produces more light than the Cherenkov effect.

The limitation on the range of explorable energy sparked the idea of a more ambitious and unprecedented experiment: measuring solar neutrinos at very low energies, below 1 MeV. Raghavan and I began involving leaders from other groups, including Frank Calaprice from Princeton University in the US and Franz von Feilitzsch from the Technical University of Munich in Germany, along with physicists from Milan, Pavia, and Genoa. Several American physicists also joined, including Stuart Freedman, renowned for his quantum mechanics experiments addressing unresolved questions, and Martin Deutsch, who discovered positronium, an unstable system of an electron and its antiparticle, the positron. By the spring of 1988, discussions were in full swing, continuing into the summer of 1989, with meetings alternating between Italy and the US, where I was still working at Fermilab.

The goal of measuring low-energy solar neutrinos meant capturing the majority of neutrinos reaching Earth, as the flux increases exponentially at lower energies. The proposal evolved into replacing Borex with an experiment capable of detecting solar neutrinos down to 150 keV. This smaller-scale experiment was aptly named **Borexino** (Italian for "small Borex").

The main challenge of such an experiment was suppressing natural radioactivity, which emits particles up to 5 MeV. Given the extremely low probability of neutrino interactions with matter, radioactivity had to be reduced to unprecedented levels. Calculations revealed the scintillator would require a radio-purity of no more than 10^{-16} g of contaminants per gram of pure substance—equivalent to 1 g of contaminants among 10 millions of billions grams of pure material.

I vividly recall a pivotal meeting in January 1989 at Argonne National Laboratory, where Freedman was working. That morning, I drove from Fermilab with Laura Perasso, a physicist from the Milan group working with me at the time. Heavy snow began falling in the afternoon, and I worried we wouldn't make it back. Raghavan brought a plexiglass model of the detector featuring a honeycomb design with multiple small scintillator-filled cells—a concept requiring substantial materials and proving impractical. During this meeting, we decisively opted for a single vessel to contain the scintillator, a decision Martin Deutsch also strongly supported. By the summer of 1989, we finalized the concept for Borexino: an "onion-like" detector with the scintillator at its core, surrounded by layers of progressively higher radiopurity from the outside inward.

I advocated for installing the experiment at the underground Gran Sasso National Laboratory (LNGS), operated by the Italian National Institute for Nuclear Physics (INFN) in the Abruzzo Apennines (Fig. 3.3). This proposal was quickly embraced by the team, as Gran Sasso offered far superior facilities compared to other underground laboratories in the US, Canada, or Japan.

After this decision, some physicists abandoned the project, considering it too difficult or even, for some, impossible. I still remember the letter Friedman sent me at the time, expressing his admiration for my courage in deciding to proceed with the experiment. He admitted he did not feel up to it, citing the extreme complexity of the endeavor and the practical impossibility of securing funding in the United States.

As a result of my explanations and discussions, INFN, particularly its president at the time, Nicola Cabibbo, showed great understanding and support for the project. Thanks to this, we secured the necessary funding to proceed with the Research and Development (R&D) phase. German agencies also contributed to the effort.

Fig. 3.3 The Rome-L'Aquila-Teramo motorway enters a 12-km tunnel. Approximately midway through the tunnel, beneath the Corno Grande—the prominent rock dihedral visible in the photo—the Gran Sasso underground laboratory is situated on the right lane. This site is shielded by 1400 m of rock, equivalent to about 4000 m of water, providing exceptional conditions for scientific research. (*Source: Gran Sasso laboratory*)

This marked the beginning of 1990.

For this experiment, I had to partially rebuild my team in Milan. Nine members of the Milan group who had worked with me at Fermilab chose to stay there to repeat the successful experiment we had recently completed, aiming to increase the statistical accuracy of the results, that is, the number of interactions that could be obtained and therefore analyzed. However, I was not particularly interested in this continuation, as in my opinion the higher statistics would only strengthen existing measurements without revealing anything fundamentally new. As a result, only four physicists from the Milan group at Fermilab joined the new experiment. Unfortunately, a highly skilled electronics technician and Franco Manfredi, a leading electronics physicist, decided not to participate, believing that the electronics work for Borexino was less engaging.

Consequently, I had to recruit new physicists and engineers. I also managed to involve research groups from the universities and INFN divisions in Pavia, Genoa, and Perugia, in addition to Princeton University and the Technical University of Munich, already collaborating and of course the Milan group. These collaborations developed further as the project progressed.

Before concluding this section, it is worth discussing the four nuclear reactions in the Sun that produce neutrinos, which I have mentioned earlier. These are the proton-proton (pp), Beryllium-7, pep, and Boron-8 reactions.

The proton-proton fusion reaction (pp), the progenitor of the solar chain, emits a flux of neutrinos with a maximum energy of approximately 0.4 MeV. In contrast, the neutrinos from the Beryllium-7 and pep reactions are monoenergetic, with energies of 0.862 MeV and 1.5 MeV, respectively. The neutrino flux from the pp reaction is much higher than that of the Beryllium-7 and pep reactions. Finally, the Boron-8 reaction produces neutrinos with energies extending up to about 16 MeV, but its flux is significantly lower than the others. Another reaction, referred to as hep, produces a flux of neutrinos that is definitively too small.

As I already explained if we were to focus solely on neutrinos with energies above 5 MeV the neutrino flux constitutes roughly 1/1000th of the total solar neutrino flux.

To a better and a deeper understanding you can visit the Annex 3.1.

Annex 3.1

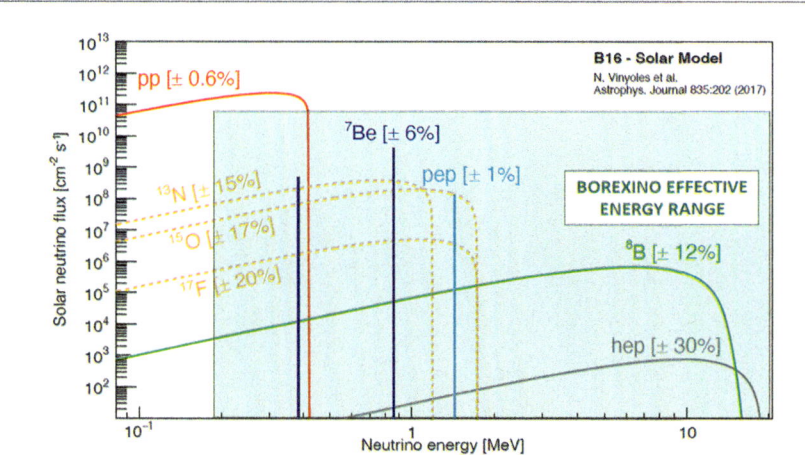

Energy of Solar Neutrinos: on the horizontal axis, the energies of the neutrinos are represented, while the vertical axis shows the frequency of the various energy values. It is important to note that the vertical axis uses a logarithmic scale, meaning each successive major division represents a tenfold increase.

This scale highlights that the frequency of low-energy neutrinos is vastly greater than that of high-energy neutrinos. The peak frequency occurs at energies below 1 MeV.

4

The Battle for Unprecedented Radiopurity

4.1 The Start Up

The first major challenge of this experiment was achieving an unprecedented level of radiopurity—a feat met with widespread skepticism. Many believed such purity was unattainable. Natural radioactivity pervades all materials, air, and water, posing a significant hurdle.

Take air, for instance: it contains radon gas, a decay product of radium in the uranium decay chain. In enclosed spaces, especially underground, radon can accumulate to dangerous levels as it escapes from radium-bearing rocks. Radon emits alpha particles—helium nuclei—that, while harmless to the skin (it poses serious risks if inhaled). Water, too, presents challenges. Spring water, such as that found beneath Gran Sasso, becomes radioactive as it flows through rocks containing radioactive isotopes. Even the materials used to construct the detector posed contamination risks. However, the main challenge was to minimize the radioactive isotopes in the core of the experiment: the scintillator.

The scintillator, a material that emits light when particles interact with it, was the most important part of the detector. This light emission, triggered by energy release, was central to detecting neutrinos. The scintillator emits photons—quanta of energy carried by light (which is an electromagnetic wave)—when particles transfer their energy, generating measurable signals. Quanta are grains of energy.

In the early 1990s, we launched a research and development program to pioneer methods for achieving the necessary radiopurity in the liquid scintillator. This task was met with immediate obstacles, including limited support from Gran Sasso's laboratory staff. At the time, the laboratory was still in its infancy and lacked experience supporting complex experiments. Faced with these difficulties, we chose to work independently. My goal was to create a self-sufficient team of physicists and engineers capable of managing all aspects of the experiment autonomously. I often joked that we needed to function like an "independent republic," relying on the laboratory only for regulatory and safety compliance.

Overcoming technical challenges became a long-term endeavor. For example, delays in laboratory services often felt insurmountable. When a pump failed, the laboratory staff would call an external company, which could take at least 1 week to respond. Fortunately, our collaborator Corrado Salvo, a physicist and chemist with technical expertise, frequently stepped in to resolve such issues quickly, bypassing delays. This independent approach, coupled with my persistence, sometimes strained relationships with laboratory engineers. I frequently called them—even on Sundays—to address unresolved problems. One winter Sunday, for instance, the heating system failed because fuel had not been ordered in time, leaving us without heat for a few days.

A more critical issue was the frequent failure of the uninterruptible power supply (UPS) during power outages. These outages caused voltage surges that damaged photomultipliers—our "electronic eyes." After many setbacks, we installed a battery system capable of sustaining power for 2 days, partially solving the problem.

Despite these challenges, our determination to overcome unprecedented obstacles kept us motivated. Achieving extreme radiopurity required designing a specialized detector to test our progress. Existing plasma-source mass spectrometers, with a sensitivity of one part per ten billion (10^{-10} g/g), were far from sufficient. We needed a sensitivity of one part per ten million billion (10^{-16} g/g). This seven-order-of-magnitude gap led us to develop the Counting Test Facility (CTF), a groundbreaking instrument capable of validating our efforts. While constructing the CTF delayed the main experiment, it was essential to demonstrate feasibility to funding agencies in Italy (INFN), Germany, and the United States, but also to us before starting the experiment.

Convincing young physicists to commit to this long-term project was another challenge. Many were apprehensive about the potential delays to their careers. I encouraged them by emphasizing the opportunity to contribute to groundbreaking technical advancements. Some were inspired by the unique challenges of the experiment; others were already captivated by its audacity.

Taking on this responsibility alongside co-leaders Calaprice and von Feilitsch was challenging, but my belief in the project's potential and my intuitive "physical sense"—the ability to gauge feasibility before detailed analysis—drove me forward. While intuition can sometimes falter, rigorous experimental controls complemented it and proved invaluable in this case.

Looking back, I recognize the risks taken by these young physicists. Yet, without risk, great achievements are impossible. Their courage and dedication were instrumental in realizing this extraordinary scientific endeavor. **I started serving as a spokesperson and essentially as the person in charge of the experiment at the beginning of the research and development phase in 1990.**

4.2 The "Mission Impossible" Truly Begins

Amidst a series of challenges, we embarked on developing innovative methods and specialized instruments for the radio-purification of the scintillator. At the heart of this effort was a close collaboration with the Princeton group, including Frank Calaprice, Bruce Vogelar, graduate students, and technicians. A key contributor was Princeton chemical engineer Jay Benziger, who pioneered new purification methods and refined existing ones in his laboratory. Additionally, Raju Raghavan partnered with a young researcher from Pavia, conducting related studies at the Bell Labs chemistry laboratory for several months.

While perfecting purification methods in a small, room-sized lab, was a vital first step, scaling up to purify 1000 m^3 of scintillator presented an entirely new set of challenges. Large-scale purification demanded substantial facilities, specialized containers, special transport systems, and complex equipment. Even carefully selected materials could release radioactive isotopes, making it critical to prevent contamination of the scintillator. The larger the purification system, the greater the risk, as more material came into contact with the scintillator. To address this, we focused on developing and rigorously testing purification methods in the laboratory before adapting them for industrial-scale application.

Over three and a half years, we refined and validated a range of techniques, ultimately establishing four primary purification methods: ultrafiltration, low-temperature distillation, water extraction, and nitrogen stripping. These methods worked in concert, with the second and the third operating alternately and the others simultaneously.

Here's a summary of these techniques:

- Ultrafiltration: Removed fine particulates with filtration down to 10 μm (10 microns correspond to 10 millionth meter)
- Low-Temperature Distillation: Operated at reduced temperatures to minimize the extraction of contaminants embedded in the steel surfaces of distillation columns.
- Water Extraction: Leveraged the immiscibility of certain contaminants to separate them into a water phase.
- Nitrogen Stripping: Eliminated gaseous impurities by introducing high-purity nitrogen gas into a purification column. The gas, free from radioactive isotopes, flowed upward while the scintillator entered from above. Nitrogen was purified using a cryogenic system with activated carbon (Fig. 4.1).

These methods were developed and implemented despite numerous secondary challenges:

Fig. 4.1 Detail of the stripping column

- Corrosive Properties: The scintillator components were corrosive, particularly to iron, necessitating the use of stainless steel containers or iron coated with inert polymers like Teflon.
- Surface Treatment: All contact surfaces required meticulous treatment to remove embedded impurities, followed by polishing to a mirror finish. This ensured ease of cleaning and the removal of dust and particulates.
- Specialized Operators: Personnel working at Gran Sasso needed specialized training in radiopurity or expertise in handling highly sensitive materials.

Every detail of the experiment required a custom, ad hoc approach, reflecting the unparalleled precision and attention to detail demanded by Borexino.

To validate these methods, we relied on the Counting Test Facility (CTF), a crucial benchmark instrument. The CTF allowed us to rigorously test and refine each purification technique, ensuring their effectiveness and suitability for Borexino's stringent requirements.

This painstaking process, involving innovative thinking and exhaustive testing, underscored the experiment's uniqueness and the dedication of everyone involved. Borexino was not just a scientific endeavor—it was a test of ingenuity, resilience, and the willingness to venture into uncharted territory.

4.3 Are We Capable of Winning the Battle of Radiopurity?

The construction and installation of the Counting Test Facility (CTF) proceeded in parallel with the development of scintillator purification methods. During this time, the collaboration grew significantly, bringing together a diverse and talented group of researchers.

Two young physicists from the University of Munich, Lothar Oberauer and Stefan Schoenert, joined the effort, supported by Curie grants that enabled them to spend a couple of years at Gran Sasso. Frank Calaprice from Princeton also took 1 year of leave of absence to dedicate himself fully to the work in the underground laboratory. The Milan team expanded as well, with Gioacchino Ranucci, Marco Giammarchi, Emanuela Meroni, Laura Perasso, and Barbara Caccianiga joining the effort. Caccianiga, a newly graduated physicist, had worked with me on an experiment at Fermilab and returned from the U.S. to join our group in Milan. Two additional grant holders joined, including a promising young woman who had written her thesis on Borexino. Her potential was evident, and I successfully secured a permanent position for her at INFN. However, personal challenges disrupted her trajectory when she experienced a heartbreak after her boyfriend left her. These moments reminded me that, in addition to overseeing the scientific aspects of a large experiment, guiding young researchers often requires navigating the complexities of their personal lives.

Meanwhile, new collaborations enriched the project. A group from Pavia, led by Giorgio Cecchet, and a Genoa group, coordinated by Giulio Manuzio, joined the team. A team of chemists from the University of Perugia, specializing in scintillator chemistry, also became involved. On an international level, I reached out to the Joint Institute for Nuclear Research (JINR) in Dubna, Russia. My earlier collaboration with JINR on a CERN-supported experiment at the Serpukhov accelerator in the 1970s facilitated this connection. Oleg Zaimidoroga responded enthusiastically and sent a young physicist, Oleg Smirnov, to join the team. Smirnov began his work in Milan before transitioning to Gran Sasso, where he set up a test system for photomultipliers. Over the course of the experiment, Smirnov became an integral contributor across multiple areas.

I spent much of my time at Gran Sasso during this period but made it a priority to return to my family every weekend. Having endured long absences during previous projects at CERN, Serpukhov, and Fermilab, I understood the strain this placed on my wife and family. Fortunately, Gran Sasso was relatively close—an hour and a half by car to Rome, followed by an hour-long flight to Milan. The balance between work and family life is one of the greatest challenges for physicists in nuclear and subnuclear research, where work often requires long stints at distant laboratories. Only with the support and understanding of courageous spouses can such a demanding career coexist with a fulfilling family life.

The CTF was conceived as a scaled-down, simplified version of Borexino. It consisted of 4 tons of scintillator, identical to what would later be used in Borexino, housed in a thin nylon balloon. This was surrounded by 1000 tons of highly purified water and monitored by 100 photomultipliers (electronic eyes) capable of capturing the faint light produced in the scintillator, amplifying the signals, and converting them into electronic pulses (Figs. 4.2 and 4.3). The construction of this test detector was highly challenging, as we were only beginning to grasp the extent of precautions and techniques required to achieve record-breaking levels of radiopurity. Every step demanded ingenuity and meticulous attention to detail, as we worked to build a detector unlike anything that had been attempted before.

The CTF became a vital testing ground, allowing us to refine our methods and prepare for the ambitious goals of Borexino, one step at a time. The

Fig. 4.2 The external tank of the CTF, a large cylindrical structure, is shown installed in Hall C of the underground Gran Sasso Laboratory. Positioned at the center of the hall, the tank features various control and monitoring systems on its upper section. Among these is a continuous recording system for measuring radon levels in the hall's atmosphere. Additional systems visible in the photo are auxiliary components essential for operating the CTF. These are not described here, as delving into their technical details might overwhelm the reader. Hall C itself is an impressive space: 30 m wide, 30 m high, 100 m long. It is lined with aluminum panels and equipped with a large, movable scaffolding system. This scaffolding, combined with an overhead crane, facilitates the transportation of heavy equipment throughout the hall

construction and installation of the CTF was done in parallel with the development of purification methods.

The development and installation of the Counting Test Facility (CTF) involved collaboration across multiple institutions, disciplines, and individuals, each contributing expertise to overcome the technical and logistical challenges. The photomultipliers, a crucial component of the detector, were developed in collaboration with the British company EMI. To achieve the required low levels of radioactivity, special materials such as glass, ceramics, and a dynode cascade with good radiopurity were used, considering that these materials typically contain non-negligible radioactivity. Gioacchino Ranucci, our engineer, dedicated months to working closely with EMI to refine these components.

Fig. 4.3 The internal structure of the CTF photographed by a camera inside the external tank. You can see the nylon container of the scintillator, the structure with the photomultipliers mounted and the relative cables. This is the cover of a 1994 issue of the scientific journal Science. (*Source: Science, License number OP-00165285, License date 20 December 2024*). This image was also featured on the cover of the German magazine Natur Wissenschaften in 1997 and the Hungarian magazine Természet Világa in 1996

The construction of a purification plant for highly purified water was another essential aspect. This facility incorporated four advanced systems, including molecular-level filtration and a stripping column where water was introduced from above and ultrapure nitrogen from below. The system design and construction were spearheaded by Marco Giammarchi and our versatile technician, Roberto Scardaoni (Fig. 4.4).

The collaboration with Princeton University was instrumental during this period. The nylon vessels, later used in Borexino, were fabricated at Princeton. Additionally, a water extraction plant was designed to purify the CTF scintillator, a task carried out by Jay Benziger, Bruce Vogelar, Fred Loeser, and others from Princeton. Steve Kidner, also from Princeton, played a key role at Gran Sasso. He relocated to Italy with his wife, settling in the picturesque town of Assergi near the laboratory. Steve's enthusiasm for the area's beauty, friendly

4 The Battle for Unprecedented Radiopurity 51

Fig. 4.4 Water radio-purification plant

people, and excellent food made his transition seamless. As a jack-of-all-trades, he was invaluable, particularly for the components managed by the American groups. Notably, the 4-ton scintillator container was built at Princeton and funded through INFN, with Princeton University effectively acting as a contractor.

The data acquisition software was developed by Alexey Golubchikov, a talented Russian computer scientist from Dubna. INFN provided him with a 5-year contract, the only means to support his stay in Italy since Dubna could not fund staff abroad. Alexey quickly mastered the Italian language and eventually settled in Italy with his family, finding work in Rome after his INFN contract ended. Alexey collaborated closely with Danilo Giugni, an engineer from Milan who designed and built the CTF's internal components. Their partnership was critical in ensuring the functionality and reliability of the detector.

Another key contributor was Istvan Manno, a Hungarian physicist sent to Italy by a colleague of mine from my time at the École Normale Supérieure in France. Istvan worked with us from 1989 to 1997, supported first by INFN contracts and later by International Affairs Funds (FAI). He developed the reconstruction and analysis code for the CTF data, which later became the foundation for Borexino's data processing software. Istvan's family accompanied him to Milan, where his son attended Italian schools and became fluent in Italian. His son's integration was so deep that he became a passionate supporter of one of Milan's soccer teams, even returning to the city for important matches after the family moved back to Budapest. Istvan's son later became an official of the Hungarian Ministry of Foreign Affairs also because of his knowledge of Italian, English and French, learned at the Italian school.

The electronics for the CTF, including process controls, photomultiplier monitoring, and related software, were overseen by Gioacchino Ranucci, Emanuela Meroni, and Gemma Testera. These three researchers, who began working with me in 1990, remained integral to the project through its conclusion. Gioacchino was my right-hand man throughout the experiment, while Gemma, a recent graduate from Genoa, won a research position at INFN and took charge of hardware and software development.

The work environment at Gran Sasso fostered a sense of camaraderie, particularly among those who worked directly on-site. I recall a later episode involving Gemma, who usually drove to Gran Sasso. One day, she had a minor car accident, which caused considerable worry, especially as she was pregnant at the time. Thankfully, she was unharmed, and the incident highlighted the strong bonds within our team. The CTF era exemplified the teamwork, resilience, and technical ingenuity required to achieve the unprecedented levels of precision and radiopurity that would pave the way for Borexino.

At that time, financing was a complex challenge. We received funding from the INFN and the German colleagues obtained funding from their own national agencies. They constructed the CTF external containment cylinder using carbon steel coated with a resin (Permatex). This resin protected the walls by isolating them from the polar molecules of ultrapure water. Water molecules, as electric dipoles, are ellipsoidal in shape. At one end, the two hydrogen atoms lack electrons, resulting in a positive charge, while the oxygen atom at the other end carries two extra electrons, creating a negative charge. (Atoms with an excess or deficiency of electrons, and therefore electrically charged, are called ions). When water is purified to remove ionic contaminants that would otherwise bind to one end of the molecules, the ultrapure water molecules can infiltrate between the iron atoms, posing unique material challenges.

Another issue I faced was providing salaries for the Russians who came to Italy to work with us, as well as for other Italians essential to the project. As the experiment progressed, managing these financial and logistical demands became increasingly difficult.

We spent most of our days in the underground hall, emerging only for meals at the canteen and, in the evenings, for dinner and sleep. It was somewhat alienating not to see the sun all day, but we were so engrossed in the work that we hardly noticed. I recall Frank working at the top of the CTF, always listening to opera music on a radio.

I took on multiple roles, acting as both the project spokesperson and a quasi-project manager. In hindsight, this was a mistake because it was extremely difficult to oversee everything while continuously making technical and scientific decisions. Nonetheless, I did my best to keep everything coordinated.

4.4 The Moment of Truth

The Counting Test Facility (CTF) began data collection in the latter half of 1994 after being filled with scintillator purified through water extraction, ultrafiltration and nitrogen stripping. The internal vessel had been filled with purified water from our water purification system. The filling process with scintillator began in mid-December 1993 and extended through January 1994. It was carefully synchronized: the scintillator was introduced into the nylon vessel while purified water, treated through the water purification system, was simultaneously added outside. This maintained identical levels on both sides, preventing any rupture of the delicate nylon container.

When procuring the primary scintillator component for the CTF, we had not specified strict requirements for radiopurity. I vividly recall the arrival of the first tanker—its condition clearly demonstrated that no special precautions had been taken. It was an ordinary chemical transport. Concerned, I questioned the driver about the tanker's cleaning process. Reassuringly but misguidedly, he proceeded to purge it with nitrogen for safety, a step that did nothing to enhance the purity of the liquid. Nevertheless, we managed to achieve sufficient purification using the water extraction method. As explained earlier, this technique exploits the polar nature of water molecules to attract and bind impurities, effectively removing contaminants.

For the scintillator filling, the Princeton team was setting up a liquid-handling system with pipes to connect the detector to storage and purification vessels. However, this setup required additional time, which I was eager to avoid as delays were mounting. After discussions with various collaborators, the Munich group stepped in to help expedite the process. Lothar Oberauer and Tanja Hagner arrived at Gran Sasso with a modestly sized Teflon pump and tubing. Teflon, being virtually free of radioactive nuclides, was an excellent choice. They connected this system to the nylon vessel inside the CTF, and the setup was ready in record time.

The filling took place during the Christmas and New Year holiday period, a time when many colleagues were away with their families. Bruce Vogelaar, however, stayed on-site, bringing along his uncle and father-in-law. His dedication during this time was extraordinary, and I remain deeply grateful for his commitment.

By January 1994, we began measurements and found traces of natural radioactivity from Uranium and Thorium families within the project requirements, indicating that our purification methods were effective. These two radioactive decay chains are among the most significant contributors to natural radioactivity, so their suppression was a major success. However, we detected an unexpected excess of counts, suggesting the presence of other sources of radiation in the scintillator. Initially, this anomaly did not alarm us; we assumed it would be easy to identify and resolve.

In March 1995, I presented a seminar at the laboratory in the presence of INFN President Luciano Maiani, the board of directors, and other stakeholders. I shared our data and results, which demonstrated the feasibility of the project. Despite lingering skepticism, Maiani showed remarkable courage and approved funding for Borexino, committing himself alongside us to the project's uncertain outcome. Franz von Feilitsch further reassured us by securing additional financial support from German agencies.

Frank, who had flown in from the United States the night before the seminar, nearly missed the event due to tight scheduling. We dispatched the laboratory's car, an Alfa Romeo, to pick him up at Rome's airport. The skilled driver navigated the route at high speed, avoiding traffic police and potential fines, ensuring Frank's timely arrival, who said that he had taken two jets: one from US to Rome, and a second one from Rome to the laboratory. When he joined us that evening, we had dinner together. Surprisingly, despite the positive outcome, Frank seemed a bit unhappy. When I asked why, he admitted he didn't know—he felt he should be happy but wasn't. This reflected his naturally cautious and somewhat pessimistic outlook, which often contrasted with my own persistent optimism.

Approving an experiment at INFN is a complex process that requires more than just the president's endorsement. Proposals must pass through scientific committees, particularly Scientific Committee No. 2, which oversees experiments in "passive physics" (those conducted without accelerators). For Borexino, this meant resolving the issue of approximately 500 excess counts in the detector before approval could be granted. Without addressing this anomaly, the committee would have rejected our proposal outright.

The collaboration began an intense period of discussions and analyses, with countless meetings and debates. Hypotheses were proposed, yet none offered clear evidence. I remained confident that the purification system was functioning properly, a stance that earned me sharp criticism, particularly from Martin Deutsch, who dismissed my views as unscientific.

Gioacchino Ranucci, however, hypothesized that the excess counts were due to external radiation rather than an issue with the scintillator itself. In December 1995, during a heavy snowfall that blanketed the fields and roads around the laboratory, Ranucci and I decided to conduct a critical test. We repeatedly opened a valve connecting the 1000-ton water shield surrounding the detector to the Hall C environment, introducing doses of radon gas into the water. Radon, abundant in the underground air, provided a controlled way to study the effect of external radiation on the detector. We recorded data for about 4 h after each radon release, alternating shifts through the night. While one of us rested on a cot, the other monitored the electronics and data acquisition system. During one of these quiet moments, I wondered whether radon in the water shield could diffuse through the nylon membrane into the scintillator. After calculations, we concluded that radon diffusion through the nylon was extremely slow and posed no significant contamination risk.

That night proved decisive. Analyzing the data, we realized that the excess counts were indeed caused by external radiation penetrating the nylon vessel and reaching the scintillator. Further investigation identified the source: the Permatex resin coating the internal wall of the iron tank. While it protected the tank from the highly purified water, it emitted radiation, albeit at a low level.

After an additional month of tests and data interpretation, we confirmed that the necessary purity levels for the experiment had been achieved. This breakthrough meant that the experiment was feasible. In September 1996, the INFN committee officially approved Borexino!

The success of the CTF brought not only scientific satisfaction but also political challenges. By then, it had become the most sensitive detector at tons level in the world. During an interview at Gran Sasso, Frank Hartman, an American collaborator and retired US Navy submariner, was asked whether the detector could identify a nuclear-powered ship entering the Adriatic Sea. Although the correct answer was an unequivocal "no" (it would have been nearly impossible to detect a few neutrino interactions amid solar neutrino events), Hartman replied affirmatively. This statement appeared in *Corriere della Sera*, one of Italy's most widely read newspapers. The article sparked controversy, prompting the secretary of the Green Party to question—either in Parliament or the Senate—whether the Gran Sasso laboratory had been constructed for military purposes. The Ministry of University and Research contacted the INFN president, who urgently requested a detailed report from me to clarify the matter. Although the issue eventually subsided, it served as an early warning of future unfounded accusations from environmentalists.

Some friends asked me how time passes during long nights of data collection. The experience is different in underground laboratories compared to those hosting particle accelerators. At accelerator facilities, experiments are assigned fixed time slots—50, 100, 200 h or more—requiring tightly coordinated day-and-night shifts to maximize efficiency. In contrast, experiments relying on continuous emissions from the Sun or cosmic sources lack such scheduling constraints.

During nights in the underground lab, the hum of equipment fills the silence, creating a unique atmosphere for reflection during breaks between operations. For me, these moments carried a blend of emotions. I felt a naive yet profound sense of contributing to humanity's collective knowledge, adding small pieces to our understanding of the universe. Yet, thoughts of my distant family often intruded, accompanied by doubts about whether the sacrifices—mine and theirs—were worth the price. In the quiet of the underground, with fewer distractions, these reflections became sharper. The balance between personal sacrifice and scientific ambition is never easy, and those nights were a constant reminder of the weight of both.

4.5 The Counting Test Facility Is Being Upgraded for Further Important Tests

The CTF served as a critical benchmark for Borexino. Following the experiment's approval and the invaluable experience gained with the CTF, we turned our attention to addressing a significant challenge: preventing external radon and other contaminants from increasing false signals in the scintillator. One potential solution involved installing an additional nylon balloon around the scintillator container to absorb external radiation—a concept we needed to test using the CTF.

During this period, we discovered that some photomultipliers (PMTs) in the CTF were malfunctioning. To investigate the issue, we hired two professional divers from Genoa to dive into the detector's water, dismount, and recover the two problematic photomultipliers. The analysis revealed that their seals were defective, compromising their performance. As a result, we decided to drain the detector and replace the faulty photomultipliers with new ones featuring improved sealing. To ensure this, Paolo Lombardi, an engineer from

Fig. 4.5 Photomultiplier assembly in CTF 2. An operator is checking the supports and the cable connections

the Polytechnic of Milan working with us, and Augusto Brigatti, one of our specialized technicians, spent 2 months at EMI, the company that had collaborated with us to develop the photomultipliers. They set up a dedicated laboratory for sealing the PMTs and trained the company's employees in the sealing process.

This initiative was not only aimed at resolving the immediate issue with the CTF but also served as preparation for Borexino, which required the production and sealing of 2212 photomultipliers. This proactive approach ensured both the reliability of the CTF and the scalability of our solutions for the larger experiment.

First, we prepared 100 photomultipliers (PMTs) with the newly improved sealing to be installed in the CTF (Fig. 4.5) At the same time, we took the opportunity to replace the optical concentrators with smaller, more efficiently designed ones, optimizing their ability to direct photons toward the PMTs. Additionally, we installed another nylon balloon, referred to as the *shroud* or *outer vessel*, around the scintillator container (Fig. 4.6). This marked the transition to what we called CTF2.

The tests conducted after the second filling of the detector were highly successful. Encouraged by these results, we decided to incorporate a suitably scaled shroud into the Borexino design, further enhancing its ability to mitigate external contaminants and radiation.

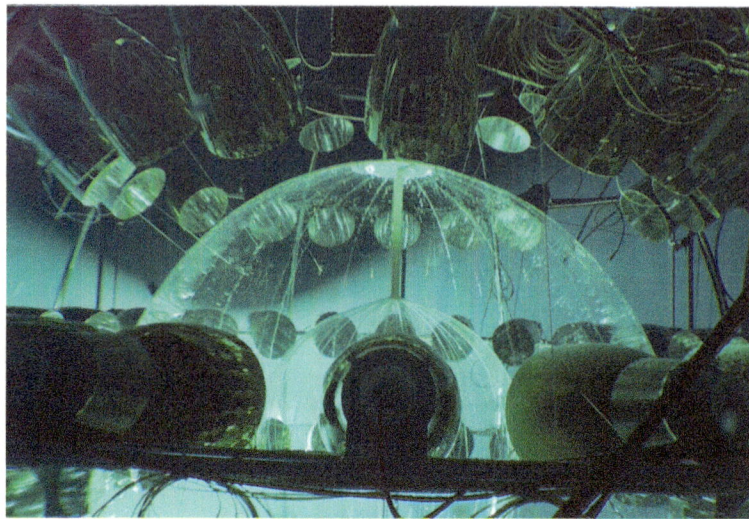

Fig. 4.6 The CTF2 in operation. Clearly visible are the nylon container holding the scintillator and the outer shroud, which acts as a shield against external radioactive emissions. The photomultipliers are equipped with optical concentrators—structures with mirrored aluminum walls shaped to deflect light hitting the walls directly onto the photomultiplier, maximizing photon collection efficiency

5

First Phase of Detector Construction

5.1 The Mission Impossible Begins

After the results obtained with the CTF and the support of INFN for the experiment, we enthusiastically began our work. We also received financial backing from German agencies, while the American support arrived only at the end of 1997 or early 1998.

Physics experiments are generally complex and require advanced technologies. It is crucial that the researchers who design and build them possess a variety of skills. This experiment, in particular, was at least partially interdisciplinary, involving physics, astrophysics, chemistry, and, also for structural challenges, engineering. My priority was to form a team comprising physicists, engineers, and individuals with strong chemistry expertise. The engineers' expertise was vital in both the design and construction phases also because the Gran Sasso area is classified as a seismic zone. During the construction phase, after resolving various challenges and finalizing the detailed project, it was crucial to adhere to the established plans. While unforeseen issues may arise, changes should be avoided unless absolutely necessary, as last-minute solutions might not be fully thought. Engineers were essential in maintaining this discipline, as the creative thinking of physicists—so vital during the design phase—could become counterproductive or even hazardous during construction. The engineers collaborating with us were from the Milan Polytechnic University, particularly from the nuclear engineering degree program. They brought a combination of engineering expertise and in-depth knowledge of nuclear and subnuclear physics.

In the meantime, the laboratory's management had changed. The new director, Sandro Bettini, a colleague from Padua, requested a revision of all laboratory infrastructure to align with the seismic conditions of the area. He also required the presence of a GLIMOS (Group Leader in Matters of Safety).

Bettini sought a more realistic seismic profile than that adopted by the Abruzzo region, which he deemed overly optimistic. To address this, I approached the Milan Polytechnic and enlisted engineer Ezio Faccioli, a European expert in seismic studies. Faccioli developed a seismic profile for the area, reflecting past earthquakes in L'Aquila. This profile was officially adopted by the laboratory and, unfortunately, proved accurate: the 2009 L'Aquila heavy earthquake closely matched his predictions.

We also needed an engineer specializing in construction in seismic zones. At the Milan Polytechnic, I found engineer Alberto Castellani, head of the seismic construction department. Castellani evaluated the Sphere and Water Tank (described later). While the Sphere was already built to withstand potential earthquakes, he calculated and implemented external reinforcements for the Water Tank, which are still clearly visible today.

I have a vivid memory of the collaboration we had with Alberto Castellani and the deep respect I held for him—not only for his expertise but also for a particular episode I wish to recount. The laboratory had commissioned a company to calculate the seismic resistance of the entire detector. At our request, Eng. Castellani was also asked to perform his own analysis. The company's findings differed from Castellani's, and they firmly defended their results as the most accurate. We then convened at the laboratory with Castellani and representatives from the company. During the discussion, Castellani asked me for a sheet of paper, took a pen from his jacket pocket, and began writing out the equations related to the problem. Step by step, he demonstrated where the company's engineers had overestimated their calculations. By the end of his explanation, they acknowledged Castellani's solution as correct.

This scene has stayed with me, reinforcing my belief that, at all levels, before turning to computers or similar tools, one must first have a clear understanding of the problem, a vision of its solution, and the mathematical framework to translate it into calculations.

Our primary goal was to develop a detailed project for the detector and its auxiliary systems. Achieving the required radiopurity for the scintillator was a significant milestone, but many challenges remained. The entire detector had to be constructed to preserve this radiopurity. Every component and material had to be carefully selected and custom-built. The scintillator's central vessel also required shielding from radiation originating from surrounding rocks and the underground hall environment.

Although the 1400 m of rock above the laboratory shielded most cosmic rays, neutrinos passed through unhindered, and a small number of muons (μ), a particle belonging to the family of the electron and neutrinos which I mentioned in previous chapters, still reached the detector. These muons, approximately six particles per square meter every 5 h, posed additional challenges.

Structure of the Detector The detector's design is based on a spherical onion-like structure, where radiopurity increases towards the center. The innermost layer, the scintillator vessel, is surrounded by multiple protective layers (see Fig. 5.1).

1. **Inner Vessel (IV):**

    - A spherical nylon container, 125 μm thick and 8.5 m in diameter, holding 280 tons of pseudocumene-based scintillator. The nylon thickness is so thin in order to reduce the amount of material in contact with the scintillator and, consequently, the emissions from radioactive nuclei present in the nylon.

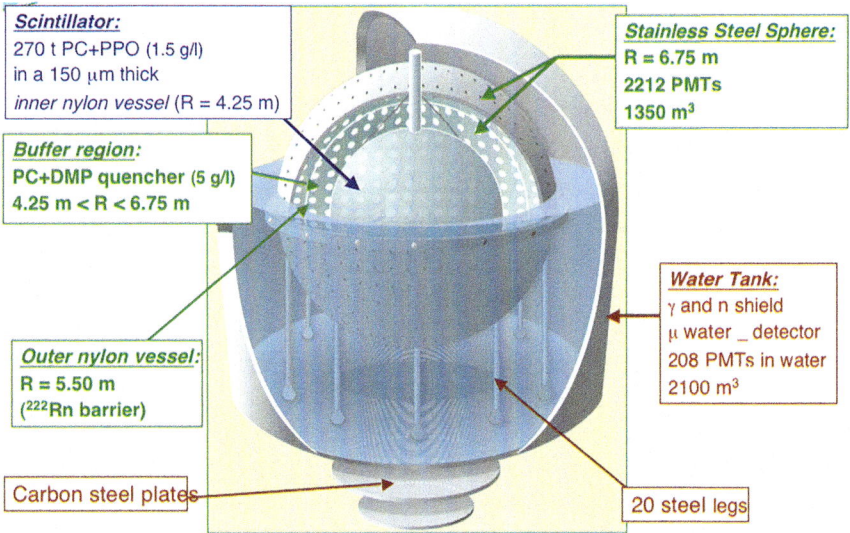

Fig. 5.1 Schematic representation of the Borexino detector, accompanied by brief descriptions of its various components. At the center is an inner nylon vessel- IV- containing 270 tons of scintillator material, PC+PPO, with a radius of 4.25 meters. Surrounding this is a buffer region filled with a PC+DMP quencher, extending from 4.25 to 6.75 meters. Enclosing these is an outer nylon vessel-OV- with a radius of 5.50 meters, serving as a radon barrier. The entire assembly is housed within a stainless steel sphere- SSS- with a radius of 6.75 meters, containing 2212 PMTs and a volume of 1350 cubic meters. This sphere is submerged in a water tank- WT-, which acts as a gamma and neutron shield, with 208 PMTs in 2100 cubic meters of water. The structure is supported by 20 steel legs and rests on carbon steel plates

- The scintillator: an organic solvent, Pseudocumene (PC), derived from petroleum, is free of benzene and non-carcinogenic and a solute (PPO) at 1.5 g/L, which enhances photon emission and adjusts the emitted light's frequency for better detection by photomultipliers.

2. **Outer Vessel (OV):**

- An 11-m-diameter nylon balloon surrounding the IV, acting as a barrier against emissions from surrounding components.

3. **Stainless Steel Sphere (SSS):**

- A 13.5-m-diameter sphere housing the IV and OV, and supporting 2250 photomultipliers.
- The space between the IV and SSS contains 1000 tons of buffer liquid (pseudocumene with added the DMP, a quencher for the low photon emission from PC) to shield against radiation from the photomultipliers, the steel of SSS and WT, and external sources.

4. **Water Tank (WT):**

- The entire assembly is immersed in 2100 tons of highly purified water within an 18-m-high, 20-m-diameter tank.
- The water shields against rock emissions and external radiation. Additionally, 208 photomultipliers monitor muons passing through the tank.

In Figs. 5.2 and 5.3, the Borexino logo is reproduced, along with a photograph of the collaboration around 1998, when we were finalizing the layout of the detector.

During this period of intense work, one of the many journalists visiting the Gran Sasso laboratory asked me if our plan was to build a large detector following the CTF. I replied that we were designing the Borexino detector. The journalist then remarked that we should also consider the aesthetics of the detector because, in his view, science and art share the same purpose. I responded that our priority was to address the technical and scientific challenges first and that the detector would ultimately be constructed in the most suitable manner to achieve our goals.

The discussion continued for a while, though I participated only minimally, as I was preoccupied with more pressing matters. Many years later,

Fig. 5.2 Logo of the Borexino experiment. The Sun is depicted with tongues of fire, it emits electron neutrinos, which may arrive at Gran Sasso as electron neutrinos, muon neutrinos (μ-neutrinos), or tau neutrinos (τ-neutrinos)

Fig. 5.3 The Borexino collaboration around 1998. Standing from the left there are Frank Calaprice, Oleg Zaimidoroga, Gerd Heusser, Raju Raghavan and Jay Benziger, while from the right, still standing, are Emanuela Meroni and Silvia Bonetti; on the right behind Meroni, with the yellow sweater, is visible Gioacchino Ranucci. On the knees the second from the left is me, and the third is Marco Giammarchi

when we were well advanced in collecting data and processing results, that conversation came to mind as I read an article in a magazine. Reflecting on it, I believe the journalist had expressed the concept inaccurately. The only connection between art and science lies in their shared interest in describing reality, though their methods and approaches are vastly different. Art conveys an individual's interpretation of reality, while science seeks to represent reality in an objective way through answers derived from experiments and observations.

That said, some aspects of the CTF, for instance, could indeed be seen as having an aesthetic quality.

When we began building the detector, a discussion arose with the Munich group, primarily during a meeting in Budapest organized by Manno, followed by a second meeting in a mountain refuge in Bavaria, and ultimately concluding at the University of Munich, where the final decision was made.

The debate centered on a proposal to increase the size of the Inner Vessel and to change the scintillator. Franz suggested using a liquid material called PXE, which had the advantage of a much higher flash point compared to pseudocumene (44 °C for pseudocumene, meaning it could only ignite above that temperature). However, the Princeton group disagreed with the proposal, and we in Milan, as the swing vote, decided against these changes—particularly regarding the scintillator. After years of studying pseudocumene's properties and its potential for radio-purification in the CTF, starting from scratch with a new material seemed unreasonable.

Nonetheless, we agreed with the Munich group to test PXE in the CTF. This test, conducted mainly by Stephan Schoenert and Tania Hagner from the Munich group, involved a different purification method using Silicagel, which I will not delve into here. The results showed that PXE purification in the CTF yielded worse outcomes than pseudocumene for Thorium, Uranium, and especially Carbon-14, along with other inferior properties that would take too long to explain in detail.

Ultimately, we maintained our original decision for the detector.

5.2 A Long Task: The Construction of the Detector

The construction of the detector took a long time, from autumn 1995 to April 2007. However, as I will explain in the following paragraphs, we were halted for two and a half years due to a court decision in Teramo.

The start of construction was quite difficult: funding was initially clear only from the INFN, with German agencies clarifying their support in 1996, while American funding arrived toward the end of 1997-beginning 1998. After several visits with US colleagues to try to convince the US National Science Foundation (NSF) to support Borexino, I asked Luciano Maiani, president of the INFN, to accompany me on a visit to the NSF headquarters in Washington. Maiani agreed, as there were ongoing contacts between INFN and NSF, particularly because they were jointly supporting other experiments. The visit was successful, and Frank Calaprice received the first grant at the end of 1997. I was later told that John Bahcall, father of the Standard Solar Model (already considered the reference for solar neutrino experiments), had also intervened in writing.

I believed that collaboration should be expanded. So, I visited the group at the University of Heidelberg, Till Kirteen's home institution, which had carried out the Gallex experiment at Gran Sasso, mentioned earlier. The group agreed to join Borexino and took responsibility for addressing all issues related to radon, which is present in both air and water. They were well-equipped to measure extremely low levels of radon and other contaminants, and they were experienced in purifying nitrogen, which was necessary for stripping. Additionally, Heidelberg offered Neutron Activation analysis, an extremely effective method for studying contaminants in materials. The most experienced member in this area was Gerd Heusser, who sent his pupil, Mathias Laubenstein, to Gran Sasso. Laubenstein remained there permanently and conducted highly sensitive routine measurements.

I also visited Paris to meet a group that had previously been at the Collège de France and later moved to the University of Paris 7. There was significant interest in our experiment among the French colleagues, but unfortunately, a national decision had been made not to fund Borexino in order to concentrate resources on a French submarine experiment on neutrinos off the coast of Marseille. Despite this, the french colleagues collaborated on a limited basis, using university funding, and were present at Gran Sasso.

A little later, the collaboration expanded further with the inclusion of a group from the Russian Kurchatov Institute and a Polish group from Krakow. Together with the previous groups from Princeton, Munich, Milan, Pavia, Genoa, and Perugia, we finally formed a collaboration strong enough to take on the construction of the detector. And so, the adventure began!

Shortly after the start of this phase of the experiment, Rudolf Mössbauer, Nobel Prize winner for the effect bearing his name and, in a way, head of all physics activities at the University of Munich, requested a meeting with all the senior members of the various groups at the Gran Sasso laboratory. After an initial discussion, they asked me to leave the room so they could discuss my position in the experiment, where I had until then served as spokesman and coordinator. Mössbauer did not agree with my being the spokesperson. I must acknowledge the loyalty of Franz von Feilitzsch, head of the Munich group, who argued that no one else could fulfill this role, which I had effectively managed despite the funding challenges, which arrived in stages. In the end, my role as spokesperson was confirmed, which essentially meant I was responsible for directing and coordinating the work of the collaboration.

The first step was to hold a meeting every Monday afternoon with everyone working permanently at Gran Sasso on the experiment. The goal was to establish the weekly program and coordinate the work of all participants. However, implementing the weekly plan was difficult from the outset because group leaders often gave instructions to their team members present at Gran Sasso, regardless of what had been agreed upon in the meeting or my decisions.

The construction of the detector was very complex because every material, component, and method had to be selected with great attention to detail to avoid contaminating the radiopurity of the scintillator. Nothing in Borexino is standard!

We started with selecting the steel to be used for the sphere SSS and the water tank WT. Steel samples were measured in the low-activity laboratory installed underground at Gran Sasso by Mathias Laubenstein. The measurements were primarily done using very high-purity Germanium detectors. Once we had selected the acceptable steel, we could begin constructing the main structures, such as the SSS sphere (Fig. 5.4) and the large WT (Figs. 5.5, 5.6, and 5.7).

Fig. 5.4 The stainless-steel sphere (SSS) of the detector, with a diameter of 13.5 m (for comparison, the average height of a room in a flat is 2.7 m). It contains a total of 1300 tons of liquid and supports 2240 photomultipliers. In this image, it is in the process of installation. The holes visible on the sphere's surface are where the terminal parts of the photomultiplier are inserted for connection to the power and signal cables

5 First Phase of Detector Construction

Fig. 5.5 The external tank (water tank-WT) of the detector during installation. The steel sphere-SSS- is mounted inside. The reinforcements are installed for static in the event of earthquake, following the Castellani instructions. This photo was featured on the cover of the April 1999 issue of Europhysics News, the magazine of the European Physical Society

Another concern was the photomultipliers. At the time, the leading manufacturer was the Japanese company Hamamatsu, which had supplied the large photomultipliers for the SuperKamiokande experiment. However, we could not use commercial photomultipliers due to concerns about radio purity. As already said for CTF, Gioacchino Ranucci took charge of this task and succeeded in finding a relatively small company in the UK: EMI. This company agreed to work with us to develop custom photomultipliers. The result was 8-inch photomultipliers made from glass, ceramics, and other materials carefully selected for their low radioactivity, which have been used also in Borexino. To avoid contaminating the glass during manufacturing, special crucibles were used. These, however, were prone to cracking when exposed to the high temperatures needed to shape the glass. There were problems with the company because these crucibles kept breaking, and we had to bear the additional costs.

Fig. 5.6 Again the external tank (water tank). In this photo, which shows the upper part of the tank, you can see the so-called organ pipes through which the cables of the photomultipliers mounted inside come out. There are eight pipes in total

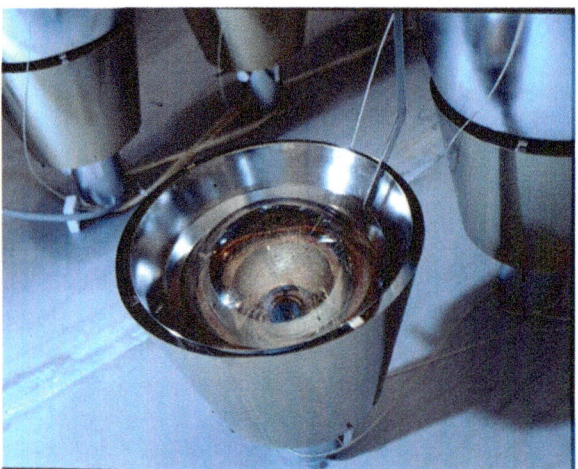

Fig. 5.7 The photomultiplier's "eye" is visible in the image. Inside the photomultiplier, a metal photocathode absorbs incoming photons and emits electrons. These electrons are then multiplied through a cascade of dynodes until they reach an anode, which generates an electrical pulse. During the electron multiplication process, the electrons can be deflected by the Earth's magnetic field. To prevent this, a cylinder made of "mu-metal" is mounted around the photomultiplier, effectively shielding it from the Earth's magnetic field

While the photomultipliers were being manufactured, they were continuously tested at Gran Sasso in the assembly hall. The testing system was organized by Oleg Smirnov, a Russian from Dubna, with significant contributions from Aldo Ianni, a graduate of the University of Perugia at the time. They conducted a substantial portion of the tests. We had to build a darkroom to ensure the photomultipliers were not exposed to light during laser tests. Additionally, we shielded the photomultipliers from the Earth's magnetic field using coils through which electric current flowed. These coils created a magnetic field opposite to that of the Earth, effectively canceling it out. Each coil could accommodate eight photomultipliers at a time. Finally, in the detector, we decided, for practical reasons, to use mu-metal, a magnetic material that also can cancel the Earth's magnetic field (Fig. 5.7). The photomultipliers were also equipped with so-called optical concentrators. These are shaped objects designed so that light photons reaching the inner surface are reflected onto the eye of the photomultiplier. The ones we used are made of mirror-polished aluminum to achieve maximum reflection. The addition of optical concentrators serves to increase the optical coverage of the photomultipliers; in the case of Borexino, we achieved an optical coverage of about 30% (Fig. 5.8).

The sealing of the photomultipliers was carried out by EMI staff in a laboratory that we set up at their company, the same used during the CTF 2 phase.

Fig. 5.8 Here is the same photomultiplier as in the previous figure, with the optical concentrator mounted. Optical concentrators increase the area that captures photons, because they redirect photons that hit them towards the photomultiplier's "eye." It is made of aluminum that has been purified to remove any impurities on its surface, and is also mirror-finished. This finish helps to both facilitate cleaning from dust and particles, and to ensure high efficiency in redirecting photons

The activities described so far took place directly at the Gran Sasso Laboratory, but many other parts of the detector were prepared at the various home institutions and then transported to Gran Sasso. One example is the preparation of the scintillator vessel (IV), which was carried out at Princeton University. Any material exposed to air tends to become contaminated by particulate matter, which can deposit on surfaces and make them radioactive. To prevent this contamination, all operations related to the scintillator container—starting from raw material extrusion to the preparation of 125-micron thick sheets—were performed in a clean room. Cristiano Galbiati, who was responsible for constructing the vessels, had written his thesis on Borexino in our group and had completed an equivalent of the PhD program in Milan. He then joined the Princeton group, initially on a contract, and later with increasingly better arrangements. Andrea Pocar also contributed to the construction of the IV, and he too came from our Department of Physics in Milan.

But you may ask: what exactly is a clean room? A clean room is an environment where the air is filtered to trap all particles larger than a tenth of a millimeter. The room is thoroughly cleaned with detergents and acids, and access is only allowed to individuals covered in plastic fabric specially treated to be free of dust and particles, with caps and masks worn. Clean rooms are classified by their level of cleanliness: for example, Class 100 means there can be no more than 100 particles of dust or particulate matter per cubic meter of air. There are also Class 1000 and Class 10,000 rooms, while achieving Class 10 is extremely difficult.

In some cases, such as when dealing with detector components in contact with the scintillator, like nylon sheets, the air filter alone is insufficient because Radon-222, with atomic dimensions, cannot be trapped by the filter. In these instances, a cryogenic system is used to eliminate the Radon. This was the case in the Class 100 clean room used for constructing and assembling the internal vessel (IV) of the scintillator and and the OV (Fig. 5.9).

The inner vessel - IV-and the OV were then transported to the Gran Sasso Laboratory, where they were stored in a hut with controlled humidity to prevent the nylon from drying out and becoming brittle.

In the meantime, at the laboratories in Genoa, the scintillator was studied in collaboration with researchers from Perugia to determine its properties, such as the type of light it emits, how much light it absorbs, and its general characteristics. This work was coordinated by Gemma Testera.

Additionally, a storage area for pseudocumene (PC), essential for the radiopurification operations and handling, was established. This area, constructed under the supervision of Laura Perasso in collaboration with a nearby company, consisted of four containers built following the same rigorous standards

Fig. 5.9 The clean room at Princeton, where the nylon was extruded, the material assembled, and the IV (surrounded by the OV) was constructed; it is equipped with a cryogenic system to eliminate radon. Operators wear clean uniforms and masks to isolate themselves from the environment and prevent contamination of the room. It is worth noting that human breath is one of the primary sources of contamination in clean rooms

as those used for Borexino containers. These included surface treatment and precision cleaning. The storage area was also equipped with fire safety measures, such as cooling systems for the vessels and foam extinguishing systems, and surrounded by a concrete containment basin capable of holding all the stored PC (Fig. 5.10).

The intense work done during this period turned out to be particularly tiring, largely due to issues I have previously mentioned, such as interference from group managers in daily operations. To address this, we decided to appoint a project manager for day-to-day management. The first was Paul Lamarche, hired by Princeton. However, he was unable to stay full-time at Gran Sasso because his family was in the United States, leading to his resignation after less than 2 years. He was succeeded by Lothar Oberauer, who accepted the role after being offered a 1-year contract by INFN. Finally, Gioacchino Ranucci was appointed, to the general satisfaction of the collaboration.

During this time, our relationship with the Gran Sasso staff improved somewhat, thanks also to Roberto Tartaglia, an engineer hired by the laboratory in the first half of the 1990s, who was dedicated to Borexino activities.

Fig. 5.10 General view of room C, with the storage area for pseudocumene visible in the foreground. The area contains our cylindrical containers designed to hold the pseudocumene. For safety, these containers are enclosed by a square containment wall capable of holding the entire volume of liquid in case of a rupture in one or more of the cylinders. The containment area is equipped with rapid fire-extinguishing systems and oxygen sensors. Cooling systems are installed. Above the containers to spray cold water onto the cylinders if their temperature rises for any reason. Additionally, fire-extinguishing systems are in place. Later, a blowdown system will also be installed (see later)

Roberto had spent a year at CERN and thus had some experience interacting with the staff members of an experiment. For our experiment, Roberto also checked the safety, and he sometimes had to argue with operators working in Hall C who might have overlooked his recommendations. Within the collaboration, there were a few truly unique individuals. One such person was Frank Hartman, as previously mentioned. Another was Corrado Salvo. Corrado was exceptionally brilliant, with outstanding knowledge of physics, chemistry, and computer science. He was also highly skilled and adaptable. However, his unconventional approach to work raised some concerns. For instance, he had set up a camp bed in the underground laboratory and frequently slept there, even bringing his own breakfast supplies. Corrado's interpersonal interactions were often challenging due to his direct manner. He openly voiced disagreements in the corridors, sometimes making accusations, which made him unpopular with some of the laboratory staff. He also preferred working outside standard staff hours, which created friction. For example, he wanted access to the chemistry laboratory at eight in the morning, even though it only opened at nine, leading to frequent disputes with the lab managers. Despite these challenges, Corrado was a very usefull member of our team. His flexibility and ability to adapt to unexpected situations made him a highly resourceful and important member of the collaboration.

Procurement of Pseudocumene A separate chapter in the project was the procurement of pseudocumene, a substance produced in Italy from petroleum via large distillation towers located in Sarroch, Sardinia, at a facility owned by ENI (the Italian National Hydrocarbons Agency), a multinational company operating in many countries worldwide (Fig. 5.11). This plant produced pseudocumene continuously, delivering it via a pipeline directly to tankers for transport to various customers. For ENI, we were something of an obstacle to their standard production process. We required a total of 1500–1800 tons of pseudocumene, but not in a continuous flow, as we used

Fig. 5.11 The Pseudocumene production towers in Sarroch

Fig. 5.12 An Isotank for transporting Pseudocumene. It is a container designed for transporting liquid or gaseous products, such as fuel, chemicals, cement, or tar. It must meet specific requirements, including a certain wall thickness, suitability for internal pressurization, maximum sealing, and other safety and performance standards. In our case, the isotank was also specially treated internally, similar to the other containers used for our scintillator. It underwent four different surface treatments to extract contaminants embedded in the material and to facilitate the removal of dust and particulates

four specially treated isotanks designed for high radiopurity (Fig. 5.12). Each isotank could hold just over 20 tons, meaning the company could not maintain a continuous production schedule for us.

An additional challenge was radiopurity: to minimize the presence of Carbon-14, we requested that only petroleum from very old and deep geological layers be used. Carbon-14, a radioactive isotope with a half-life of 5370 years, is commonly known for its use in dating archaeological and historical artifacts. For our experiment, however, it posed a significant problem because pseudocumene cannot be purified of Carbon-14. The solution was to source petroleum from deep, ancient deposits, where the Carbon-14 had already decayed and was no longer being redone by cosmic rays. This requirement added complexity to the process, as ENI sourced its petroleum from multiple suppliers.

5 First Phase of Detector Construction

Fig. 5.13 Pumping station built in Sarroch, at the company supplying the pseudocumene, with Roberto Scardaoni, one of our invaluable technicians

Despite these challenges, the staff at Sarroch were cooperative, pleased to tackle a task different from their usual routines. Ultimately, I succeeded in finalizing the contract. Additionally, we built a dedicated pipeline to transport the pseudocumene directly from the production columns to the edge of the company's property. At this endpoint, we constructed a pumping station to fill the isotanks. All procedures adhered strictly to the radiopurity standards required for the detector (Fig. 5.13).

5.3 Two and More Dark Years

In a complex scientific experiment, such as the one I am describing here, those in charge must dedicate a significant amount of time not only to technical-scientific problems but also to organizational management. It is not just a technical-scientific effort but also an organizational one, particularly when working with numerous collaborators from various institutes and countries, as is often the case with large and complex experiments, the so-called "Big Science." Personally, I estimate that I spent about 60% of my time by coordinating the activities of various individuals and the rest determining the best technical and scientific decisions to address the problems that arose along the way. What was truly important was that I had succeeded in assembling teams of high-level researchers and engineers who had studied the best methods and all the necessary precautions to achieve an exceptionally high level of radiopurity.

In a relatively large collaboration like that of Borexino, there are not only technical-scientific and organizational challenges but also occasional misunderstandings among members, and, at times, personality clashes. It is no coincidence that I have often said that large collaborations might benefit from having a psychologist as part of the team.

Situation for the Gran Sasso Laboratory Since its inception, the Gran Sasso Laboratory had been met with resistance from local environmentalists. This opposition stemmed from a combination of prejudice, general distrust of science and technology, and lingering resentment over the construction of the highway tunnel, during which the mountain's water table had dropped below the tunnel level. Although this did not harm drinking water or its distribution, it left a lasting negative impression. The excavation of the laboratory occurred after the highway tunnel and was unrelated to the aquifer issue. The lab was connected directly to one of the highway's lanes. When Sandro Bettini became the laboratory's director, he, along with a highly competent engineer who was the Minister of Works at the time, proposed a new idea. They suggested building a third, much smaller tunnel to directly connect the laboratory to the section of the highway towards L'Aquila. This would bypass the highway, thereby avoiding interference with both laboratory activities and public traffic. The proposed tunnel would pass above the existing one, ensuring no further impact on the aquifer. However, this project faced strong opposition from environmentalists and Green Party members, who were fundamentally against any additional tunnels, regardless of their size or purpose. During a TV talk show, where the secretary of the Green Party participated, Bettini defended the project, dismantling the criticisms with scientific arguments. Unfortunately, this escalated tensions, leading to violent attacks on the laboratory and Bettini himself. Threats of various kinds were made against him, including graffiti attributed to insurrectionist anarchists known to the judiciary and DIGOS (Division of General Investigations and Special Operations). These included threats to kill the laboratory director. The situation became so severe that Bettini had a permanent police escort, and a police car was stationed at the laboratory's management offices. Local newspapers and other media further fueled the controversy, falsely asserting that the lab was discharging toxic water.

The Borexino Spil As already explained, to ensure coordination and avoid overlaps or gaps in the work, I had established weekly meetings with everyone involved in the experiment at Gran Sasso. These meetings were typically held on Monday afternoons, following a practice I had observed with great success during my time at Fermilab near Chicago. At Fermilab, in addition to the weekly meetings, there were daily briefings during or immediately after lunch to quickly address any unforeseen issues. These meetings were invaluable. I had not implemented them at Gran Sasso, as it would have been difficult to convince people to work during lunch in the Italian setting. It was also a standing rule that during the Ferragosto period (August 15), work could only proceed if at least three operators and a technical manager were present. Unfortunately, these instructions were not followed during the week ending on August 15, 2002. Not only was no technical manager present, but the laboratory staff had been reduced to basic services. Without my knowledge, the American group, urged by Frank Calaprice, decided to proceed with their work. As an American, he did not fully grasp the Italian custom of suspending activities during this period. At the time, I was on vacation, staying by the seaside because my wife had fallen ill, and I could not leave her alone. During this period, a Canadian operator, whose name I do not think it is appropriate to mention, misinterpreted the procedure for operating a valve. In the absence of Corrado Salvo, the technical manager who had set up the system, this mistake caused approximately 30 L of pseudocumene, the solvent used in the scintillator, to spill into a local stream.

This aromatic liquid is very volatile and has a characteristic smell somewhat similar to gasoline. Although it is not carcinogenic due to the absence of benzene, the small incident triggered a disproportionate reaction from those eagerly awaiting an opportunity to criticize. It was like throwing a lit match into a room full of gasoline. Environmentalists erupted with accusations, some of which were absurd, and filed a complaint with the court of Teramo.

This outburst against the laboratory came primarily from the Teramo area, because the province of Teramo did not enjoy the same advantages as the province of L'Aquila. Many laboratory employees were from L'Aquila, having won competitive selections, and local hotels and restaurants—being much closer to the laboratory—were frequented by laboratory staff, external scientists, and technicians. Furthermore, the administration of L'Aquila was politically center-right, whereas that of Teramo was center-left, which used the situation to exchange political accusations.

What finally happened is that a magistrate in Teramo forbade us from using any liquids, including ultra-pure water produced by the water purification plant. This prohibition lasted for the first year. In the second year, access to Hall C—where we were installing the detector—was restricted except for safety reasons.

Despite these setbacks, we managed to continue preparing various parts of the experiment, both hardware and software, albeit with limited autonomy and mobility. Constant monitoring from the authorities further slowed progress, and many collaborators became demotivated, uncertain whether the experiment would be shut down permanently or resumed after years of delay.

The War Against Borexino A relentless campaign against Borexino unfolded, with ongoing attacks and accusations from local newspapers and media. Even a left-leaning national newspaper amplified the story, exaggerating the scale of the spill. The judicial police, represented by the Forest Rangers, were hostile on all occasions. For instance, a local authority, which I prefer not to mention, once collected a water sample seeping from the bottom of a tub on the floor of Hall C in the presence of the INFN and our lawyer; when our lawyer pointed out that the law required three samples—one each for the judicial officer, the INFN, and the consultants—the chief threatened to remove him if he persisted. Although the sample was illegally taken and thus inadmissible in court, the real goal was to tarnish the experiment's reputation by having local newspapers report the water as heavily contaminated. Another sample was analyzed by a regional representation in the presence of a specialized Gran Sasso laboratory technician, who confirmed no trace of pseudocumene. However, after the technician left, this representation re-ran the analysis and claimed to find traces, promptly leaking this information to the local press.

I tried to keep the collaboration united, but many American researchers—especially those early in their careers—left because they struggled to renew contracts or secure new ones without publishing for several years. The situation was somewhat better for Europeans, but the delay still posed significant challenges, particularly for young scientists. I invested considerable effort in reassuring collaborators that the experiment would resume. One memorable instance was when Raghavan invited Aldo Ianni, an essential member of our team, to join Virginia Tech and work on another experiment, LENS, which ultimately never materialized. Ianni sought my advice, and I encouraged him to remain at Gran Sasso, which he did. Frank Calaprice supported me by assisting in responding to inquiries. I was unable to be at the laboratory permanently, but fortunately, an engineer from the Polytechnic of Milan, Augusto Goretti—who had married a local woman and settled in Gran

Sasso—accepted responsibility for representing the experiment in my absence during inspections by magistrates, carabinieri (the Italian military police), consultants, and firefighters.

Letters from prominent scientists, including Nobel Prize winners, emphasized the importance of the experiment and the potential harm to science if it was not completed. However, the court remained unmoved, stating that public and environmental safety took precedence.

Pseudocumene was classified as harmful to aquatic environments but not otherwise particularly hazardous. Director Bettini, in an attempt to address concerns, even asked environmentalist critics to show evidence of any dead fish, as none had been reported. The leaked liquid had mixed with abundant percolation water, which significantly diluted it and practically eliminated any danger to aquatic life.

On a lighter note, absurd accusations also emerged. A regional environmental agency dispatched a ship to the Adriatic Sea, into which flow the streams from the Teramo side of the laboratory, to check for pseudocumene contamination, even trying to bill us for the cost. The absurdity didn't end there. A smaller experiment located downstream from our detector had a researcher, known for his flatulence, who jokingly displayed a mock warning sign about 'toxic gases' (Fig. 5.14). Instead of recognizing the humor, critics accused us of using dangerous gases. Unfortunately, rather than dismissing these accusations with the laughter they deserved, we were forced to provide formal explanations, which were met with constant skepticism.

Subsequently, Hall C's floor was completely sealed to submarine-level insulation standards. I worked tirelessly to ensure the contracted company completed this work promptly.

Scientific Competition and Further Challenges Another aspect of this challenging affair was the competition from other scientific experiments. In particular, there was the Japanese experiment KamLAND, designed to measure reactor antineutrinos over a long baseline to study neutrino oscillation parameters, study which does not require radiopurity at the level of Borexino; but the Japanese team also aimed to extend their research to solar neutrinos. As is standard in fundamental physics, we had published the results of our research, including a detailed technical description of the systems and methods we developed to achieve unprecedented levels of radiopurity. The Japanese team had even visited our facilities while they were under construction at the Gran Sasso laboratory. Our concern was that KamLAND, having started their experiment earlier and potentially adopting some of our techniques, might

Fig. 5.14 The joke poster, displayed by colleagues from another experiment, was intended to humorously tease one of their own team members who was known for suffering from flatulence

achieve results on solar neutrinos ahead of us, especially given that we were halted. However, KamLAND ultimately did not reach the level of radiopurity required for studying solar neutrinos, and thus it did not pose direct competition to Borexino. This period, however, was deeply frustrating and demoralizing. Meetings convened by the commissioner, appointed by the government, overseeing the laboratory works were especially disheartening. While we were invited to these meetings, we were not allowed to intervene—even when incorrect statements or scientifically unfounded evidence were presented. We were treated as if we were schoolchildren caught misbehaving, despite the accusers' lack of understanding of the experiment or the technical challenges we faced. I recall many such meetings in Rome, where I participated alongside around 20 engineers. These engineers either remained silent or made incorrect claims, yet our attempts to clarify or correct the record were dismissed. The chairman would often cut us off with phrases like, *"OK, OK, let's leave it"*.

The situation with the regional environmental agency was no better. For instance, when they sampled water in a manhole inside the laboratory, to check the water that eventually had gone into the stream, they took the sample from the surface. Given that pseudocumene has a lower density than water and remains on the surface, this method resulted in an extremely high percentage of pseudocumene being reported—completely unrepresentative of the actual concentration in the water.

Another absurd claim came from a court consultant who stated that Hall C's atmosphere was entirely saturated with pseudocumene vapor. Despite our explanations that pseudocumene's vapor pressure at the underground laboratory's temperature made this physically impossible—and that the liquid was always contained in sealed containers and transported through closed pipelines—these facts were ignored.

We faced these problems not only when dealing with institutions but also in our relationships with private individuals. To demonstrate the usefulness of the laboratory and show the population that we could help them, we decided to measure the presence of radon in the cellars of homes, where it tends to accumulate due to poor ventilation. From these measurements, we observed radon levels higher than those in the atmosphere, which was expected. However, after the initial measurements, people told us to stop, claiming that our equipment produced radioactivity. The terrorist alarm had reached such a level throughout the province of Teramo!

Even worse things happened: teachers began telling children not to drink the city's water because it was allegedly polluted by our laboratory! One day, a woman called the lab, accusing us of being despicable people because we were supposedly polluting the aqueduct water, while we could drink bottled mineral water. She claimed this was because we were paid by the Americans, which was not only false but also absurd.

An increase in cancer cases in the region was also fabricated and blamed on us, despite the fact that pseudocumene is not classified as a carcinogen. This claim was later debunked when statistics showed no such increase. Finally, I want to recount a conversation I had with a group of self-proclaimed environmentalists who had learned that we used nitrogen. We needed nitrogen to extract noble gases such as helium from the scintillator and water. Once processed, the nitrogen was passed through activated carbon to absorb any liquid residues and was then tested with an instrument highly sensitive to organic molecules (such as those from PCs) before being released into the atmosphere. They accused me of contaminating the atmosphere with nitrogen, and I replied that the atmosphere is made up of 75% nitrogen. Their response was, "It's the usual arrogance of science!"

Safety We had already contracted, some years earlier, the engineer Domenico Barone—specialized in safety within the petrochemical industry and occasionally a court consultant—to oversee safety measures and conduct HAZOP (Hazard and Operability Analysis) and QRA (Quantified Risk Analysis). During the years of inactivity, Barone and I prepared a comprehensive document for each part of the detector, including all auxiliary systems. These documents detailed system descriptions, operational procedures, and risk analyses. We practically filled an entire closet with these documents to ensure we were prepared for any eventuality. Additionally, we enhanced all safety devices in meticulous detail. I will highlight two examples:

1. PC Storage System
   Each storage container was equipped with a burst disk designed to rupture if the internal pressure exceeded a certain threshold, such as in the case of a fire. To prevent the gas from escaping into the room if a rupture occurred, the containers were connected to pipes that directed the gas to a tank filled with water, rendering it harmless. Fortunately, this system never needed to be activated.
2. Nitrogen Blanket System
   The detector and all vessels containing pseudocumene were maintained under a dynamic nitrogen blanket that had been purified through a cryogenic plant. Since the blanket was dynamic, the nitrogen flow passed through two activated carbon systems, which were alternately regenerated. After the activated carbons, the nitrogen flowed into a high-resolution detector for organic molecules before being released into the atmosphere.

6

Completion of Construction and Installation: The Detector Goes into Operation

6.1 The Resumption

After two and a half years of being blocked and the liquid-proofing works for Hall C, the judge declared that the situation was under control and that we could resume our activities. It was the winter of 2014. However, we were forbidden from using water for another year.

None of us were sent to trial, except for the director of the laboratory and the president of the INFN, apparently because the laboratory lacked authorization regarding water discharge. I personally had to pay my legal costs and, of course, my lawyer's fees. We were also investigated by the Italian State Court of Auditors because some money from Italian Institutions had been spent on research and tests as a consequence of the affair. Ultimately, we were completely exonerated.

This ordeal also caused problems with the laboratory staff, as some engineers had been investigated too. The staff's attitude suggested that we had brought issues into the laboratory that would not have arisen otherwise. Gradually, we succeeded in easing these bad feelings.

I must emphasize that no one—none of the environmentalists, police officers, magistrates, or local authorities involved in this affair—ever asked us about the purpose of our experiment or the science behind it!

The stoppage also had financial repercussions. During the two-and-a-half-year halt, in the annual funding discussions, the INFN referee responsible for our project asked Gioacchino Ranucci and me if we truly intended to continue. However, our firm determination left no room for objections, and

INFN continued to support Borexino. The NSF, which operates on a biennial funding system, posed no issues—Princeton simply requested a renewal once activities resumed. Unfortunately, the German funding agencies withdrew their support, and I believe our German colleagues at Gran Sasso continued their work with university funds.

When we were finally allowed to resume experiment preparation, I made significant changes to our work organization. I established an operational group dedicated to completing the detector, working independently of the group leaders. This eliminated interference from group leaders who, at times, gave conflicting instructions to their collaborators present at Gran Sasso, slowing progress and creating problems. It was not easy to get the group leaders, especially the Americans, to accept this change, but it was the only efficient way to proceed. I appointed Augusto Goretti as coordinator of the operational group. He accepted the role with determination, fully aware of the responsibility and the potential legal repercussions of even minor mistakes. The operational group included physicists, engineers, and technicians well-versed in all aspects of the experiment, particularly in addressing the challenges of creating a detector with exceptional radiopurity. However, I judged this insufficient for handling the scintillator, which was based on an aromatic liquid. Fortunately, I found two highly experienced technicians, Ambrogio Cubaiu and Fausto Soricelli, in Porto Torres, Sardinia, where a large oil refinery operated. Both had retired a couple of years earlier. Their expertise proved invaluable in preventing even the smallest spills—down to individual drops—on the laboratory floor, which could have led to unpredictable consequences.

Another decisive change was the establishment of a steering committee to oversee the day-to-day work on-site, chaired by Marco Pallavicini, a senior physicist from Genoa who effectively acted as a project manager. I believe both Augusto Goretti and Marco Pallavicini played very important roles in the success of the detector.

Once the new organization was in place, and we could resume work, we were all enthusiastic about restarting. We worked with great determination. Personally, I worked 10 h a day, and despite being tired in the evenings, I was always eager to begin again the next morning.

After the court-ordered stop and the conclusion of the investigation, our relationship with the laboratory staff improved. Augusto Goretti's work was appreciated by the engineers at Gran Sasso. Additionally, at our request, the laboratory contracted Stefano Gazzana, an engineer from Rome, to manage various tasks and act as an interface with the laboratory staff. Gazzana was a unique individual—he studied engineering and theology simultaneously at the university. I fondly recall our many discussions on a wide range of topics. On weekends, if there was no work at Gran Sasso, he would return to Rome to spend time with his family and collaborate with parishes to assist people. After the experiment, Goretti was hired by the laboratory, as his experience with Borexino was deemed invaluable. Meanwhile, Gazzana moved on to work with a physics research institute in Rome before finding another job. Unfortunately, I lost touch with him, which I regret because he was not only highly skilled in his work but also deeply interested in the study of human nature in its entirety.

For the experiment, we resumed where we had left off. Among the many tasks was the construction of clean rooms. We built five clean rooms around the detector, as every component introduced into it had to be meticulously cleaned. The largest room was at the entrance to the detector, more specifically at the WT (Fig. 6.1). There, a Class 100 tunnel was built to ensure that all photomultipliers—and, before them, all scaffolding components needed for installation in the steel sphere—underwent precision cleaning. The steel sphere SSS itself, as I have previously explained, was equipped to function as a Class 1000 clean room.

Fig. 6.1 Photomultipliers equipped with optical concentrators and encased in mu-metal, meticulously cleaned and double-bagged within the clean chamber located at the inlet of the detector water tank

All valves, pumps, connectors, and joints posed additional challenges. Not only were these components difficult to source because they required special properties, but nearly all of them also had to be modified. For instance, in valves and pumps, any part in contact with the pseudocumene or scintillator had to be made of Teflon, which is practically non-radioactive. Additionally, all valves and joints were enclosed in nitrogen-atmosphere boxes to isolate them from environmental radon underground, even in the event of a breakage. The detector itself was sealed to extremely high standards to protect against radon.

I could continue indefinitely, recounting the specialized methods and tools used in constructing and installing the detector and auxiliary systems. However, I will stop here. This detector was fundamentally different from those used at accelerators, where, apart from specific components, most equipment is built with relatively standard methods.

We started the procurement of pseudocumene, which before pumping was routinely checked by various Milan members of the collaboration in addition to Masetti from Perugia. But it didn't end there because the cosmic rays impacting on the product produced in it radioactive nuclides and then it was important that the product was shielded underground as soon as possible. So we set up an organization that allowed the loaded isotank to leave immediately Sarroch so as to embark as quickly as possible on a ferry that connected the island of Sardinia to the continent; once disembarked on the peninsula, the vehicle proceeded immediately to the Gran Sasso where it immediately went underground. In this way the time between the loading and the arrival underground was reduced to about 20–22 h.

We were been stopped for two and a half years and when I resumed contact with Eni to ask to start the Pseudocumene production, I found a new manager of the production in Italy, who at first impact told me that the contract was no longer worth because we had been stopped for a long time and we were out of time. This reaction was terrifying for me because I was aware that in Italy it was the only producer and I did not even know if there was any other company in Europe that produced Pseudocumene; I has been told by the American colleagues that a company in the United States was producing it but it would have been infinitely more complicated to start a new process with a US company. I then used all my persuasive skills invoking scientific interest and insisting that for their multinational it could be a source of pride to have helped such an important experiment, even if honestly I was not so sure that we would have had great success. This manager very seriously but also very kindly explained to me that for a company like theirs, which had to face many problems around the world including also helping the populations where they found oil or gas, supplying us in reality was a great nuisance. In the end I succeeded to persuade him and I could breathe a sigh of relief.

6.2 The Radiopurity

All these precautions allowed us to achieve a scintillator composition with only two carbon-14 nuclei per one billion times one billion carbon-12 nuclei, the latter being non-radioactive. Additionally, for the Uranium-238 family, the achieved radiopurity corresponds to the presence of just two radioactive nuclides among 100 million billion stable nuclei, which are non-radioactive. A similar achievement was made for the Thorium-232 family, with only five radioactive nuclei per billion billion stable nuclei.

Radon, which is omnipresent in underground air and water, was reduced to less than one count per day for 100 tons of pseudocumene. Similarly exceptional results were achieved for Argon and Krypton, which are present in the air.

In the end, we succeeded in attaining levels of radiopurity significantly exceeding the design requirements. Without these levels, we would never have been able to achieve the remarkable results we obtained.

6.3 The Set Up Within the Sphere

The Borexino photomultipliers, manufactured and sealed at EMI, were subsequently tested in the external laboratory at Gran Sasso. This testing took place in a completely dark room using the electronic system originally assembled during the CTF phase by Oleg Smirnov and Aldo Ianni. The gain of the photomultipliers—defined as the ratio between the number of electrons collected by the photomultiplier and the total number transmitted to the electronics—was measured at this stage.

The photomultipliers then needed to be mounted inside the SSS (stainless steel sphere), which is over 13 m in height. To prepare for this challenging task, Paolo Lombardi and Augusto Brigatti attended a mountaineering school in Liguria, at the starting point of the Alpine arch near Genoa, to learn techniques for safely securing themselves to rocky surfaces and for using ropes and harnesses effectively.

The installation process began with constructing a scaffolding structure to support the work. Following this, steel ribs, shaped to conform to the spherical geometry of the SSS, were carefully installed using ropes and

Fig. 6.2 The scaffolding during the installation of the photomultipliers. Constructing the scaffolding was a significant undertaking, as every component had to be meticulously cleaned with acid in a clean room, double-bagged, and subsequently introduced into the SSS, which was itself equipped as a Class 1000 clean room

mountaineering harnesses. This intricate work required a combination of precision, expertise, and physical skill.

The scaffolding floors were subsequently installed, supported by the steel ribs, to provide a platform for operators during the installation process (Fig. 6.2). However, the scaffolding could only reach a certain height near the dome of the sphere. For the final installations at the top, a cherry picker was employed to complete the task (Fig. 6.3).

The installation of the photomultipliers, which required meticulous precision and care, was carried out by Augusto Brigatti, Paolo Lombardi, a technician from the Gran Sasso laboratory, and another from Milan, Massimo Orsini and Sergio Parmeggiani, along with a couple of collaborators from the Kurchatov Institute in Russia. The team dedicated themselves fully to this delicate task, completing it over the course of approximately 2 months (Figs. 6.4, 6.5, 6.6, and 6.7).

Although the gain of the photomultipliers—measured before their assembly—was carefully calibrated, it requires continuous monitoring during operation to account for potential changes. In Borexino, this ongoing monitoring was achieved using optical fibers that delivered calibrated laser pulses directly

6 Completion of Construction and Installation: The Detector Goes... 89

Fig. 6.3 Installation of Photomultipliers using a Cherry Picker

Fig. 6.4 Start of the scaffolding dismounting

Fig. 6.5 Again dismount of the scaffolding

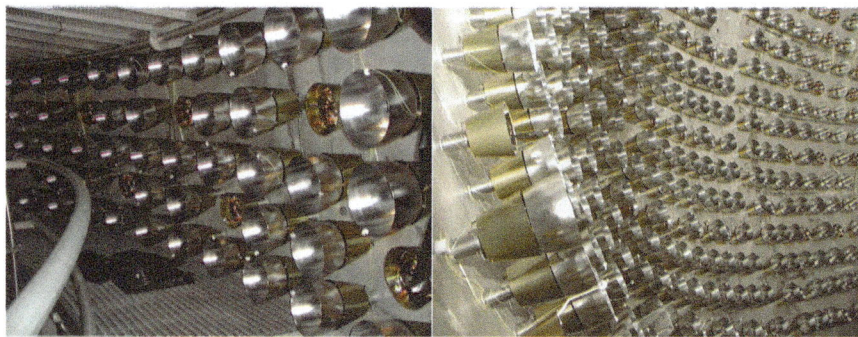

Fig. 6.6 On the left, a close-up of the photomultipliers installed with the scaffolding still in place; on the right, a section of the sphere with the photomultipliers fully installed. (The photo on the right was reproduced on the 2014 commemorative stamp of the Italian Post Office)

to the photomultiplier "eyes." This system was developed by Barbara Caccianiga, in collaboration with Bruce Vogelar from Virginia Tech, and was installed by several team members, including Lino Miramonti from the Milan group, Paolo Lombardi, and others. The timing information was crucial for pinpointing the location in the scintillator where a neutrino interaction occurred. This relies on differences in the response times of the photomultipliers: those closest to the interaction point respond first, while those farther away respond later due to the longer photon travel time. To ensure accuracy,

6 Completion of Construction and Installation: The Detector Goes...

Fig. 6.7 The dome of the sphere with the photomultipliers mounted

Fig. 6.8 Lino Miramonti is connecting the optical fibers to the individual photomultipliers for continuous gain monitoring. The fibers transmit light pulses, generated by a connected laser, to the photomultipliers

it was vital to verify that the response times of the photomultipliers were either uniform or well-characterized if they differed. José Maneira, a young Portuguese physicist and PhD student at the University of Milan under my supervision, organized and managed this continuous timing control system (Figs. 6.8 and 6.9).

Fig. 6.9 Paolo Lombardi is addressing an issue with an optical fiber that was not properly transmitting the laser pulse to the photomultiplier

Fig. 6.10 An operator, positioned outside the sphere, is testing the airtightness of the terminal section of the photomultipliers, which pass through the sphere and extend into the water tank. The seal must be 100% airtight; if air and vapors cannot pass through, we can be confident that liquids, such as pseudocumene and water, will also be effectively sealed. This test was performed on all the photomultipliers using a portable mass spectrometer

The photomultiplier terminals passing through the holes in the sphere are connected to power and data cables. Outside the sphere, the seal around the connectors is rigorously tested for a perfect seal, close to 100%. This sealing was tested using both air and vapor, with the verification conducted using a portable mass spectrometer (Fig. 6.10). In Fig. 6.11 a view of the interior of the sphere after the installation of the phototubes was finally completed.

Fig. 6.11 A view of the sphere with the photomultipliers installation completed

Before installing the inner vessel in the sphere, we performed a precision cleaning using a specially designed cleaning module. This cleaning process was supervised by Corrado Salvo, with assistance from Aldo Ianni and Lino Miramonti. The precision cleaning involved a sequence of three acids—glycolic, citric, and formic—followed by a wash with highly purified water produced by our own facility for approximately 24 h. This thorough cleaning covered all systems, including the lines transporting pseudocumene, the vessels, and any other components that would come into contact with it. The purified water used for the final rinse was produced by our water purification system, which initially had a flow rate of 1 m^3/h. Corrado increased the production to 2 m^3/h, as the rinsing process required a greater flow.

I would like to share two episodes that reflect Corrado Salvo's character. Since some people in the laboratory were skeptical about the use of these acids, Corrado took a small amount of citric acid, diluted it, and drank it to demonstrate that there was no danger. To ensure the water's high level of purification, we measured its resistivity—its resistance to the passage of electric current, which flows through any impurities present in the water. As impurities are removed, the resistivity of the water increases. Naturally, as the final cleaning process proceeded, the water accumulated some impurities and its resistivity decreased. Aldo performed a complex calculation to determine the resistivity based on the density of the impurities removed, which took a considerable amount of time. Corrado, on the other hand, did the same calculation in his head, using a simplified approach, and arrived at a result that was very close to Aldo's, in a fraction of the time.

6.4 Installation of Nylon Container for Scintillator: Inner Vessel—IV

The journalists were highly interested in this experiment, and they arrived at Gran Sasso asking all kinds of questions. I interrupt the description of the installation here to recount one of the many episodes involving journalists. Just when we were overwhelmed with work and the problems that arose, a journalist came to the laboratory looking for me. His opening question was, "Professor, is it true that we are children of the stars? There's even a song by an Italian singer that says we are children of the stars. I ask you this because you are so interested in studying the stars."

At first, I thought he was making fun of me or that the question was related to some publicity stunt organized by the laboratory's public relations office. During downtime, if I recall correctly, the office had arranged readings of poems, theatrical performances, and even a dance troupe in the underground laboratory, all to raise awareness about the existence of this laboratory of excellence.

I was a bit puzzled by the journalist's question and replied that I didn't fully understand what he meant. At his insistence, I promised I would think about it—mostly just to get him out of my way. That evening, while dining at a restaurant near the laboratory, I reflected on his question and came up with an answer. The next day, at the entrance to the laboratory, I found the same journalist, who was surprisingly waiting for me, as if his question was of great importance. I explained to him that in the primordial universe, carbon did not yet exist, even if recent observations suggest that traces of carbon appeared about one billion years after the Big Bang. A star's life cycle involves the fusion of hydrogen into heavier nuclei, including helium. Helium, together with beryllium, can give rise to stable carbon-12. In the final stages of its life, a star releases many of the elements it produced, including carbon. Therefore, carbon, which is essential for life as we know it, originates from the stars. The journalist later published an article in a local newspaper, reporting what he had seen and heard at the laboratory, including my answer. I was quite embarrassed, as my explanation, though accurate, was presented as something new, when in fact it was well-known among astrophysicists.

Returning to the installation process, once the photomultipliers were fully assembled, it was time to install the Inner Vessel. As a reminder, there are two nylon balloons in the system, similar to those used in the CTF: the innermost one, called the Inner Vessel (IV), designed to house the scintillator, while the outermost one, the Outer Vessel (OV) or Shroud, serves to protect the Inner

Vessel and its contents from contaminants originating from the photomultipliers and the steel sphere. The sphere itself functions as a clean room, though it is not Radon-free, as the filters installed did not have the necessary cryogenic apparatus to stop this gas. Radon has difficulty embedding in hard metals like steel, but it is much more likely to embed in softer materials such as nylon.

The two nylon balloons, IV and OV, preassembled inside one another, were enclosed in a nylon cover, which had to be removed during installation. To prevent Radon contamination, one option was to use synthetic air, which is generally purer than atmospheric air. By storing the air in cylinders for a few months, it would be free of Radon, which has a half-life of 3.8 days. As Radon decays, it transforms into Bismuth, which then decays into Polonium—both radioactive elements, but with much lower frequencies than Radon. However, we realized that it was impractical to fill the entire internal space of the steel sphere (SSS) with synthetic air due to its large volume. We decided on the following solution: during the installation, the Inner Vessel was surrounded by the Shroud, which was kept tightly sealed to prevent Radon from penetrating. The contamination of the Shroud's external surface was less critical since it would not come into contact with the scintillator. After the installation both nylon balloons were filled with synthetic air—preferred over nitrogen because it would not pose a danger to the operators in the event of a leak (Figs. 6.12 and 6.13).

Fig. 6.12 During installation of Inner Vessel and shroud (OV)

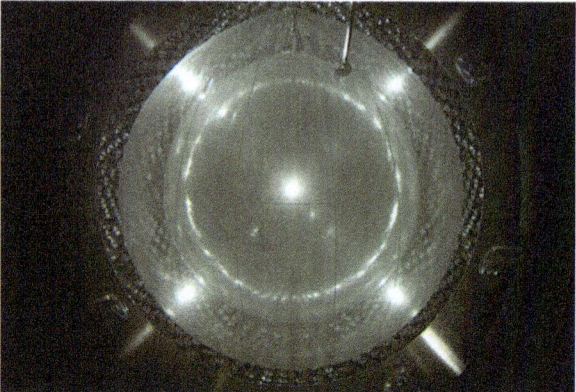

Fig. 6.13 The Inner Vessel and the Outer Vessel-Shroud after installation, inflated with synthetic air

During the installation, a small leak was discovered in the Inner Vessel, which Paolo Lombardi and Augusto Brigatti repaired using scaffolding and creating a small access window in the Outer Vessel. Inside the sphere, Bruce Vogelar installed cameras to document the status of the installation, while multiple temperature probes were placed at various points in the detector.

6.5 Inside the Water Tank

We were in the midst of our work when a young blonde woman knocked on my office door. She introduced herself as a Slovakian physicist interested in joining the Borexino project. During our conversation, I discovered she had previously worked in Italy at the Frascati Laboratory, where an electron accelerator ring was in operation (this stay her fluent Italian), and in Switzerland on medium-energy experiments. She also shared an intriguing detail: she had earned PhDs in both physics and geology. Given that Borexino included plans to study geo-neutrinos—antineutrinos originating from the Earth interior—it was evident that someone with expertise in both physics and geology would be an invaluable addition to our team. Her credentials, combined with her apparent enthusiasm and active demeanor, convinced me to accept her without further inquiries. Her name was Livia Ludhova, and she went on to play a pivotal role in the analysis and physical interpretation of data related to both solar neutrinos and geo-neutrinos. Livia became a leading figure in the study

6 Completion of Construction and Installation: The Detector Goes...

Fig. 6.14 Here, the lower part of the Water Tank is shown during the installation of the 200 photomultipliers, with three visible in the image. These photomultipliers are designed to capture photons produced by the Cherenkov effect in the tank water. Notably, they do not measure the energy released; their sole purpose is to signal the passage of a μ particle

of Earth-originating neutrinos and remains a sought-after speaker at international conferences on the subject.

Returning to the detector: outside the steel sphere, 200 photomultipliers were installed to detect the passage of muons (μ particles)/Fig. 6.14). As mentioned earlier, a small percentage of these particles pass through the laboratory's overburden. Being electrically charged and highly energetic, they generate Cherenkov light in the water of the Water Tank, which the photomultipliers process to signal their presence. Detecting these signals is crucial, as muons can mimic neutrino interactions. This installation, handled by the Munich group, employed different sealing techniques compared to the inner photomultipliers.

Since the Cherenkov light emitted in the water is far weaker than the light produced in the scintillator for the same energy loss, it was essential to minimize light loss. To address this, the outside of the sphere and the inside walls of the Water Tank were lined with a highly reflective synthetic fabric—Tyvek—which significantly enhanced light reflection (Fig. 6.15).

Fig. 6.15 The outer surface of the sphere and the inner walls of the Water Tank are covered with Tyvek. This covering is carefully designed to leave the photomultiplier sensors (eyes) exposed, allowing them to protrude through the Tyvek layer. The operator visible in the image is Paolo Lombardi

6.6 What Happens to the Data Collected by the Photomultipliers?

The output of the photomultipliers is an electrical pulse. You might wonder: how can scientific results be derived from a simple electrical pulse?

The process begins with the electrical pulse traveling through cables to the Counting Room, which houses the processing electronics and computers (Fig. 6.16). These cables connect to the photomultipliers via the so-called *anode* using specially modified submarine connectors. These connectors pass through designated holes in the steel sphere. Altogether, the cables span a total length of 124 km and weigh approximately 21 tons. After exiting the sphere, the cables traverse the water in the Water Tank and pass through the organ pipes. Outside the tank, they must be meticulously arranged—a task carried out by Paolo Lombardi and Augusto Brigatti (Figs. 6.17 and 6.18).

The process electronics was designed by Sandro Vitale from Genoa, manufactured by a company in Milan, and installed by a team that included Massimo Orsini and two Hungarian second-level engineers, George Korga and Laslo Papp, recommended by Manno. These two engineers proved invaluable, as the electronics had been installed at the beginning of the 2000s and needed to remain operational until 2021, the conclusion of the experiment. There were, of course, occasional failures over this extended period, necessitating the

Fig. 6.16 Process Electronics during Initial Installation in the Counting Room

Fig. 6.17 Paolo Lombardi and Augusto Brigatti, in safety gear and helmets, are lining up cables along the Water Tank

Fig. 6.18 The same as shown in Fig. 6.17

replacement of components that had become obsolete or were no longer available on the market. I recall George scouring the market and the warehouses of the European Centre for Nuclear Research (CERN) to secure as many components as we could find to ensure the system's continued functionality.

The collected data are processed by the electronics in the counting room. Continuous monitoring is essential, as errors can occasionally occur in this process, including malfunctioning photomultipliers that either send incorrect data or fail to transmit any data at all. In Fig. 6.19, system checks are performed in the control room, while Fig. 6.20 shows physicists leaving the laboratory late in the evening after working on these checks.

The working of the electronic process are explained in more detail in Annex 6.1.

6.7 The Filling

The first operation involved filling the Vessel-IV, the Shroud, and the SSS Sphere simultaneously with ultrapure water produced by our water purification system. This was a delicate process as it was essential to maintain the same level across all components. As already explained in the previous paragraph,

6 Completion of Construction and Installation: The Detector Goes...

Fig. 6.19 Sandra Zavatarelli from Genoa and Livia Ludhova are working in the counting room, performing quality checks on the electrical pulses processed by the electronics and the data acquisition program (as described in the text). The work was carried out in 24-h shifts, divided into three 8-h rotations. However, the curiosity sparked by the ongoing observations or the determination to complete tasks, often led many operators to spend significantly more time in the counting room than scheduled. The counting room was frequently bustling with activity, with numerous people present simultaneously. This included electronic technicians to monitor the electronics, ensuring everything functioned smoothly

Fig. 6.20 From left to right: Alessandro Razeto from Genoa, responsible for the data acquisition program, Barbara Caccianiga, and Gemma Testera, leaving the external laboratory in late evening during the checks of the data

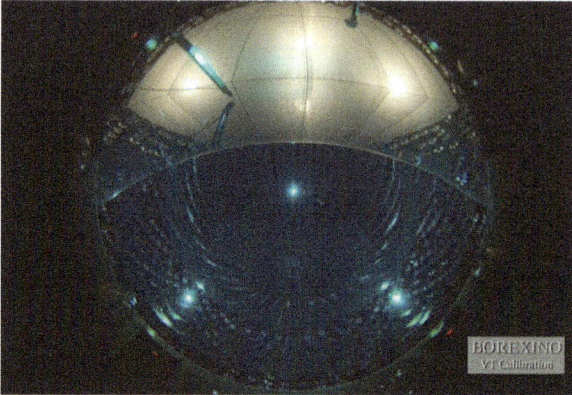

Fig. 6.21 Water is introduced from below, simultaneously filling the Inner Vessel, the Shroud, and the SSS Sphere

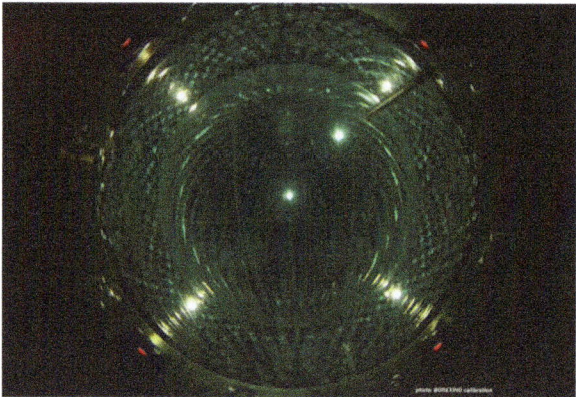

Fig. 6.22 Completion of Water Filling

any imbalance could have caused the Inner Vessel (IV), made of nylon just 25 μm thick, to break. Sensors strategically positioned in the detector governed this operation, controlling the liquid handling system. Water was introduced from below, gradually displacing the synthetic air through a valve located at the top of the IV (Figs. 6.21 and 6.22).

The systems associated with the handling of pseudocumene were carefully reviewed by Cubaiu and Sorricelli, who also collaborated on drafting procedures. These procedures were subsequently approved by both the regional firefighters and the laboratory. It is also worth noting that, because we stored pseudocumene, we were required to comply with the Seveso law—a regulation named after a small town that experienced a dioxin-related incident. This

6 Completion of Construction and Installation: The Detector Goes…

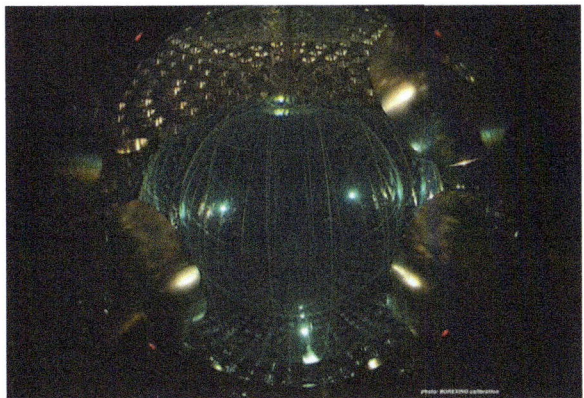

Fig. 6.23 The filling of both the Inner Vessel with the scintillator and the sphere with the buffer liquid is completed. These photographs were captured using the cameras installed inside the SSS

law mandated renewal every 5 years under the supervision and control of the fire brigade of L'Aquila.

On January 14, 2007, the first isotank carrying pseudocumene arrived from Sarroch, escorted by a police car. Subsequently, we received three isotanks per week. The deliveries typically arrived around nine in the evening, and the filling operations were conducted overnight.

The pseudocumene was processed through the skid for distillation, mixed with PPO, that had been radio-cleaned through water extraction, and transferred into the Inner Vessel. Simultaneously, pseudocumene was also sent to the sphere (buffer liquid) and mixed with DMP, which had also undergone radio-cleaning (Fig. 6.23).

Finally, the Water Tank was filled with highly purified water from our plant. The filling process began in mid-January 2007 and was completed on May 14 of the same year. With this, we could finally begin collecting data on neutrino interactions.

It had been 19 years since I first began discussing the experiment and 17 years since we started working on it! To mark the occasion, Stefano Gazzana had prepared T-shirts featuring the Borexino logo on the front and a diagram of the detector on the back. We could finally relax a little and organized a large celebration to commemorate the milestone (Figs. 6.24 and 6.25).

Cubaiu was then called back and, this time, financially supported by the Germans and Princeton to overhaul their facilities. He was still present at Gran Sasso when the devastating 2009 earthquake struck L'Aquila, which registered peak values of 7.2–7.3 on the Richter scale. Fortunately, there was

Fig. 6.24 The Borexino collaboration in 2007

Fig. 6.25 Celebration on April 20, 2007, marking the beginning of data taking. From left, after the person holding.the camera (Andrea Ianni): Viktor Machulin (Kurchatov), Laura Perasso, Giulio Manuzio, Fausto Soricelli (drinking), Ambrogio Cubaiu (viewed from behind), Surovov (Dubna, with long hair), myself (holding a glass), an to my left, Marco Pallavicini and Elena Guardincelli (Genoa)

no damage to the underground laboratory, as earthquakes typically have much less impact underground since the rock is not free but confined by other rock above and below. I remember that Cubaiu immediately conducted an inspection and informed us that everything was in order with our facilities.

Throughout this period, my role was to conduct the experiment and lead the collaboration, which included highly skilled physicists who often disagreed with one another. As a result, I frequently had to set aside individual issues, as otherwise, I wouldn't have been able to fulfill my responsibilities. This is something that makes me a bit sad, as I believe that a human being is worth more than any experiment. However, I found great consolation when Cubaiu told me that this period was the best of his life, as he had found both friends and a great family. This made me feel that I had contributed to bringing a sense of harmony within the collaboration, even amid the challenges of our work.

Annex 6.1

I would like to reiterate that neutrinos have no electrical charge and, therefore, cannot lose energy in the scintillator. However, they do collide with an electron in the atoms of the scintillator, transferring their energy—or a significant portion of it. The electron, being electrically charged, loses its energy as it moves through the scintillator, which is then converted into photons of light. These photons reach the photomultiplier's eye. The photomultiplier then converts the photons into electrons, amplifying them in the process.

Once the electrical pulse reaches the electronics, it undergoes analysis and processing (the "electronic process"). The pulse is amplified, converted into a digital signal (transforming the analog signal into a digital one), and accepted following a "trigger." The trigger accepts the pulse if certain conditions are met, with the most important being the minimum number of photomultipliers that must be hit by the photons. This number is determined according to specific criteria, which can be easily understood: even a minimal loss of energy in the scintillator must produce a sufficient number of photons to involve a certain number of photomultipliers. Below this number, the pulse could be a spurious event, such as one generated by electronic fluctuations or other phenomena.

The accepted pulses are then processed by the Data Acquisition (DAQ) system, a program that transforms the pulse into a set of input data, which are then stored in the computer's memory. These input data must correspond to the necessary parameters for the reconstruction program, which will extract the event's energy and its position in the scintillator's space. Before entering the reconstruction program, however, the processed data undergoes quality control. This step discards any data that may be problematic, such as those caused by issues in the electronics (e.g., not being properly tuned or having faulty components) or by malfunctions in the photomultipliers, which may degrade over time. After passing these checks, the data enter the reconstruction program and are then interpreted by researchers.

7

What Powers the Sun?

7.1 The Detector Operates Effectively

In May 2007, we began data collection, recording information on electrons impacted by neutrinos. Once the data were recorded, they underwent a fairly complex process, including quality control, before being written to the computer's memory. This data collection occurred around the clock, as the Sun continuously emits neutrinos in vast quantities: as mentioned earlier, approximately 60 billion neutrinos per square centimeter per second.

The organizational structure established for the detector's construction was partially modified to meet the requirements of data collection, analysis, and scientific interpretation. These tasks demand highly skilled individuals proficient in data processing, theoretical interpretation, and advanced computer applications.

I resumed the task of organizing the team: groups were formed to analyze the data and focus on specific aspects. One group worked on code to reconstruct the position and energy of the interactions, another on using neutrino energy data to identify specific nuclear reaction in the Sun via computer codes developed on purpose, and yet another on simulation code (the so-called Monte Carlo calculations). We also appointed a coordinator to oversee all these efforts, and we chose Gemma Testera for the role.

Some members of the collaboration, who had worked on other experiments during our forced hiatus, returned to dedicate themselves to data analysis. Additionally, new researchers expressed interest in joining the collaboration. In the meantime, a Russian group from the St. Petersburg Nuclear Physics

Institute in Gatchina had joined our collaboration, while other groups also increased, such as the one from Dubna, whose coordinator was Oleg Smirnov in addition of the retired Oleg Zaimidoroga.

A challenging issue arose: how to fairly acknowledge those who had been involved since the early stages and contributed significantly to the detector's installation and operation. During the 17 years of developing purification methods and constructing the detector, I had implemented a practice where individuals with key responsibilities were credited in scientific papers with a footnote describing their role, such as "project manager" or "chairperson of the operational group."

For new members, we decided they would need to wait 1 year before being eligible to co-author scientific articles. Similarly, members who left the collaboration would continue to receive authorship credit for 2 years. We also faced the question of how to address members who had contributed until the legal halt in 2002 but later moved to other experiments and published results in scientific journals. I will discuss this matter later.

The reconstruction of the energy and position of neutrino-electron interactions is critical to achieving the experiment's objectives. Energy is a key parameter because each fusion reaction produces neutrinos with specific energies, allowing us to identify the nuclear reactions based on their neutrino energy profiles. This is precisely what we accomplished. Meanwhile, the interaction position is essential for distinguishing solar neutrino signals from false events caused by contaminants.

When we began data collection and analysis, I was uncertain whether we would successfully detect solar neutrinos, as the experiment remained challenging despite achieving high levels of radiopurity. It quickly became evident that trace amounts of radioactive elements were still present in the scintillator—small enough to be considered minimal but significant enough to interfere with neutrino signals. For instance, residues of Polonium-210 and Bismuth-210 were detected, both of which fall within the energy range of neutrino signals from solar fusion reactions. These two radioactive nuclides are part of the natural decay chain of Uranium-238.

The first neutrinos we measured are produced by the nuclear fusion reaction of the Beryllium-7 nucleus plus one electron and producing Lithium-7.

As already explained, in Borexino to measure neutrinos means to measure the energy of the electrons that have been hit by neutrinos. The electrons take on part of the neutrinos kinetic energy and therefore can have various energies up to a maximum that corresponds to the total energy transported by the neutrinos. The neutrinos produced by the fusion reaction involving Beryllium-7 all have the same energy at 862 keV. When these neutrinos hit electrons they can transmit to them an energy that goes from zero to the maximum energy available to them, i.e., 862 keV. This characteristic of the electrons hit by Beryllium-7 neutrinos, i.e., having an energy that goes from zero to a maximum of 862 keV, is very specific and allows us to recognize these electrons and therefore the corresponding neutrinos, that hit them, quite easily. A complication is that Polonium-210 emits radiation that has energy precisely in this region and therefore what we need to do from a data analysis is to be able to recognize these neutrinos even if they are subjected to interference from Polonium radiation (for those who want to delve deeper I recommend looking at Annex 7.1 where the issue is better explained). The Beryllium neutrinos are disentangled by software tools from the overlap of Beryllium neutrinos and Polonium radiation.

We began the analysis with this fusion reaction because, on one hand, it has a characteristic distribution, as explained, and on the other hand, it provides the largest number of neutrino interactions, given that we cannot measure energies below 150 keV (where the majority of neutrinos from proton-proton fusion are present).

I hope this example helps the reader understand how the data analysis is conducted.

We began collecting data in May 2007, and by September of the same year, we had already obtained the first energy distribution of neutrinos produced by the reaction with Beryllium-7. This result, which I presented at the TAUP (Topics in Astroparticle and Underground Physics) world conference, generated significant attention. It marked the first time that the flux of low-energy solar neutrinos was measured in real time by isolating the contribution of just one nuclear reaction occurring in the Sun, specifically at an energy below 1 MeV. At the conference held in Sendai, Japan, our result was considered the most important among everything that was presented. For further details, we were also assigned a talk in a parallel session, where it was presented by Marco Pallavicini. But it also generated excitement in other communities: when the news spread that we had succeeded to measure the neutrinos produced by Be-7, a meeting of the SNO collaboration was underway. At the announcement, an applause broke out, as I was told by one of the members of SNO, I believe Mark Chen, who had been involved in Borexino in its early years.

The result sparked much discussion between the European members of Borexino, coordinated by Gemma Testera, and the American members, led by Christian Galbiati. The primary focus of the debate was the level of uncertainty to be reported for the rate of the neutrinos interaction belonging to Be-7 reaction. After extensive analysis in the following months, it became clear that the European team's evaluation was correct.

Naturally, we were eager to publish this groundbreaking result, but a debate arose between me and Frank Calaprice regarding who should sign the first article. I believed that only those who had worked continuously on Borexino, without leaving the collaboration during the hiatus years (2002–2004), should be listed as authors. To me, this was a matter of fairness. Those who had left the collaboration to work on other experiments had the opportunity to publish results from those projects, whereas those who remained dedicated to Borexino had signed very few papers—primarily some on the CTF results, which were far less significant than the current findings.

The Americans, however, argued that everyone who had contributed to Borexino in any capacity since its inception should be included. Ultimately, my perspective prevailed for the first article. However, in the second article, which included more comprehensive data and analysis with larger statistics obtained with the analysis of more neutrino interactions, everyone's names were included. Additionally, we wrote a separate paper on the detector, in which all contributors were listed as authors.

This discussion about authorship may seem surprising to the reader, but for scientists, published articles are a critical metric for career advancement and evaluation.

7.2 Flash on the Sun's Structure

The work of an experimental physicist, and of experimental scientists in general, involves balancing the pursuit of general themes in research with solving technological challenges. In our case, this balance includes both the quest to understand how the universe works—how Sun and stars function, and how stable our Sun is—and addressing the technological obstacles without which such research would be impossible.

For example, in the case of Borexino, the results we achieved were entirely dependent on the radio-purity we were able to attain, thanks to a suite of advanced technologies and their highly specialized applications. Even for me personally, this dichotomy—between the search for broad scientific insights and the construction of the experimental conditions necessary to achieve meaningful results—has always been at the forefront. To me, being a physicist is not merely a job like any other. It stems from a profound drive to understand the meaning of the world around us. Although the answers I have found, and continue to find, represent only small pieces of the puzzle each one adds to humanity's collective knowledge.

7 What Powers the Sun?

The purpose of the data analysis effort described here is to uncover the mechanisms that allow the Sun to shine—one of the many essential conditions for life on Earth. Before detailing the results, however, it is helpful to have a general overview of the Sun. Below, I offer a brief exploration of its structure (Fig. 7.1).

The Sun's *core* is the region where fusion reactions occur, producing the energy that makes the Sun shine. The core occupies about 10% of the Sun's radius, has a density 20 times that of steel, and reaches temperatures of approximately 15 million degrees. All the thermonuclear reactions studied in our experiments occur within this core.

Surrounding the core is the radiative zone, where energy slowly radiates outward. It takes more than 170,000 years for energy to traverse this zone. Beyond the radiative zone lies the convection zone, where energy moves toward the Sun's surface through turbulent thermal columns. At this point, the temperature has decreased to around 5700°.

The Sun's atmosphere is composed of several layers, including the photosphere, chromosphere, and corona. The photosphere is the Sun's visible surface; below it, the Sun becomes opaque to visible light. The coolest region of the Sun, located about 500 km above the photosphere, has a temperature of approximately 4100°.

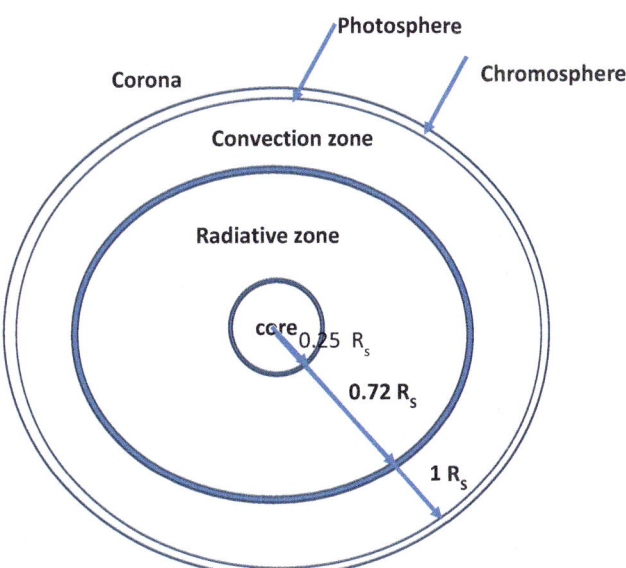

Fig. 7.1 Scheme of the Sun. R_s is the Sun radius. The innermost layer is labeled "core," followed by the "Radiative zone" extending to 0.25 solar radii, and the "Convection zone" reaching 0.72 solar radii. The outermost layers include the "Photosphere," "Chromosphere," and "Corona," with the total radius marked as 1 solar radius. Arrows indicate the boundaries between these layers

The chromosphere and the corona are much hotter than the surface of the Sun, but the reason for this remains not well understood. The chromosphere gets its name because it appears as a colored flash at the beginning and end of total solar eclipses. Its temperature gradually increases, reaching up to 20,000 °C at its peak.

The corona, on the other hand, reaches a staggering temperature of 1 million degrees, with the elements within it glowing in X-ray and extreme ultraviolet. The hot matter flowing outward from the corona is shaped by magnetic field lines into streamlined forms called coronal streams, which extend millions of miles into space.

Among all these solar layers, the most important part—and the one that concerns us the most—is the core.

The reference for understanding the Sun's composition and functioning is the *Standard Solar Model (SSM),* a widely accepted and paradigmatic framework. The father of this model is the American physicist John N. Bahcall, who developed it over nearly 50 years, working alone or with some of his pupils. The model describes the Sun's evolution from the time when the protostar was still contracting. It assumes the Sun is a sphere of gas in hydrostatic equilibrium, balanced between gravitational forces trying to compress it and the pressure gradient resisting that compression. The model also assumes that the Sun is spherically symmetric, that its core rotates slowly, and that its internal magnetic fields are approximately equal to those on its surface. The chemical composition of the Sun at its origin is considered to match that of the photosphere, comprising 71% hydrogen, 27% helium, and 2% heavier elements.

Tests of the model focus on parameters such as the Sun's mass, surface temperature, and luminosity (the solar energy reaching Earth per unit of time). An important validation of the model comes from the *helioseismology*, which examines the Sun's internal oscillations, which are primarily caused by pressure waves generated within the Sun and transported convectively toward its surface. In our context, the Standard Solar Model (SSM) is particularly significant for its predictions regarding the fluxes of solar neutrinos, their energy distribution, and the density and temperature distributions in the solar atmosphere.

The reliability of the SSM is considered very high. The solar neutrino problem arose due to discrepancies between the neutrino fluxes measured by the four experiments mentioned earlier—and the expectations of the SSM. As previously explained, this discrepancy was not caused by a defect in the model but rather by the fact that these experiments detected only one of the three types of neutrinos, namely electron neutrinos. While the Sun emits only electron neutrinos, the oscillation phenomenon causes some of these neutrinos to transform into the other two types. The problem was resolved by the Canadian experiment SNO, which studied neutrino interactions in heavy water (we have already discussed this matter earlier in Chap. 3).

The Sun is a medium-sized star, and there are many stars in the Universe of a similar size. The pp chain is the dominant energy-producing process in such stars. However, the situation is different for massive stars, which have at least 30% more mass than the Sun; in these stars the hydrogen burning occurs differently, as we will explain later. Stars form within molecular clouds—regions of relatively high-density gas in the interstellar medium. These clouds consist primarily of hydrogen, with helium accounting for 23–28%, along with traces of heavier elements. After their formation, stars spend approximately 90% of their lifetimes in a stable phase called the *main sequence*, during which they fuse hydrogen in their cores into helium under high temperatures and pressures. The duration of a star's main sequence phase depends primarily on the amount of nuclear fuel available and the rate at which it is consumed.

The Sun, currently in its main sequence, has an estimated total main-sequence lifespan of about 10 billion years. Larger stars, by contrast, consume their fuel more rapidly and thus have much shorter lifespans—ranging from a few tens to hundreds of millions of years. Conversely, smaller stars burn their hydrogen at a much slower rate, granting them lifespans of tens to hundreds of billions of years. The main sequence phase ends when the hydrogen in a star's core is entirely converted into helium through nuclear fusion. The subsequent evolution of a star depends on its mass and follows different paths based on this characteristic.

Everything discussed here about stars—their characteristics and temperatures—applies to stars in their main sequence phase b.

7.3 The Calibration

After measuring the flux of Beryllium-7 neutrinos, we continued collecting data until the autumn of 2010. During the analysis, we realized the need to calibrate the detector. But what does calibration mean? It means finding a match between the data provided by the detector and a known energy corresponding to something well-defined.

One of the major tasks in data reconstruction is estimating the energy released during interactions. To achieve this, we must accurately determine how to extract energy information from the data. An initial calibration had already been performed; otherwise, we could not have proceeded with the analysis. This initial calibration relied on the signals from residual radioactive elements, particularly Polonium-210 and Bismuth-210, which emit radiation of well-known energy. However, for a precise energy estimation and accurate reconstruction of the spatial point where the neutrino-electron interaction occurred, a more refined calibration was necessary.

One effective method involves introducing in the scintillator artificial radioactive sources that emit radiation of fixed energy. These sources deposit their energy within the scintillator, allowing us to position them at various points inside the inner vessel to verify that the energy response is uniform throughout the scintillator volume. To achieve this, we used 11 sources mounted on the end of a mobile arm connected to a rod thus allowing the sources to be positioned at various points in the scintillator. (Fig. 7.2).

The major challenge was ensuring that inserting these sources into the scintillator did not compromise its radiopurity. To address this, we installed a Class 10 cleanroom above the water tank, equipped with a glove box—a

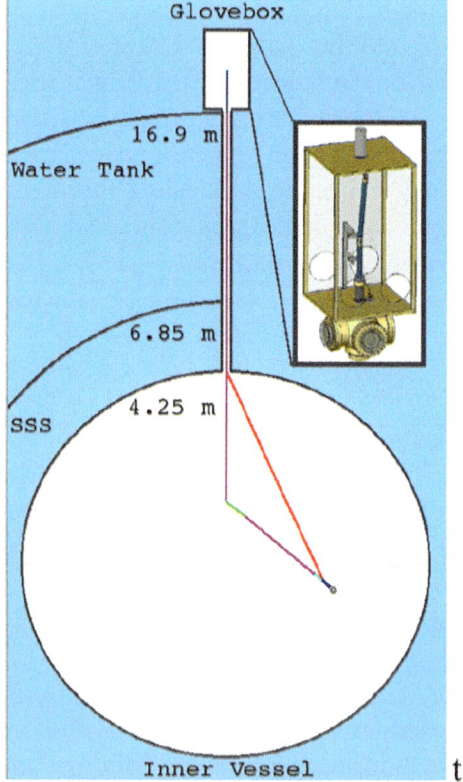

Fig. 7.2 The calibration system used for Borexino

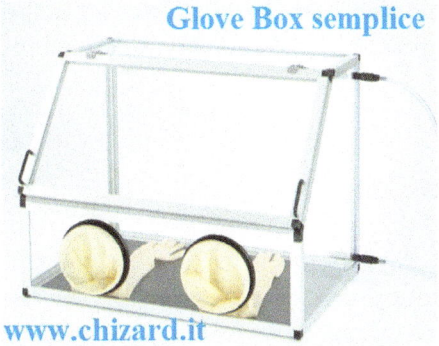

Fig. 7.3 Schematic of a glove box

sealed chamber accessible only through built-in gloves (Fig. 7.3). The glove box was continuously flushed with ultra-pure nitrogen to maintain cleanliness. Additionally, the radioactive sources were thoroughly cleaned with detergents and acids before being placed inside the box.

Despite all these precautions, a slight increase in radioactivity was observed in the scintillator at the end of the calibration operations.

The problem of using radioactive sources is always very complicated for safety reasons; they are prepared by specialized companies that deliver them closed in containers that are capable of stopping the radiation emitted and that are obviously removed at the time of use.

I remember an episode. Previously we had some need to use a radioactive sources with some difficulties due to safety issues and delivery times by specialized companies. Our colleague Oleg Zaimidoroga, head of the Dubna group working at Borexino, arrived at our physics department around noon, it was perhaps 2001, and after greetings he told that he had something for us. He then put his hand in his pants pocket and took out a radioactive source. We were all very surprised thinking that Oleg had come by plane from Moscow to Milan with a radioactive source in his pocket. I must clarify that the source emitted only neutrinos and alpha particles (helium nuclei), which are stopped by just the outermost layer of skin, in this case, human skin. Obviously the source was then put away in a safe place in the laboratory where they kept the radioactive sources and I don't remember if it was used or not. But what is remarkable is that Oleg, to our surprise, told us that he didn't see what the problem was and that in Russia they used big sources, obviously I think appropriately shielded, as a heat source in some regions of Siberia.

Thanks to the calibration, we measured the error in the reconstructed interaction position (approximately 10 cm) and in the energy estimation (approximately 1.5%).

7.4 A Second Radio-Purification

After measuring neutrinos from the fusion reaction with Beryllium, during a general collaboration meeting, Frank Calaprice remarked, *"So, we have completed the experiment!"* Initially, the experiment's goal was to measure neutrinos from Beryllium, because, as already said, these are the least difficult to detect among the pp chain reactions due to their relatively good flux and an energy of 862 keV—well above our threshold of 150 keV.

Having successfully measured the Beryllium neutrinos, we had no intention of stopping. Instead, we aimed to measure the fluxes from other nuclear reactions in the Sun, specifically the nuclear fusion of two protons and an electron (pep) and the reaction involving Boron-8. For these measurements, improved radiopurity would be highly beneficial. Toward the end of the summer of 2010, we proceeded with a second purification using methods significantly different from those employed during the initial filling of the inner vessel, which had involved distillation.

In this second purification, the inner vessel was already filled with the scintillator—both solvent and solute were mixed—making distillation impossible, as it would separate the two components. The operational group including Cubaiu, spearheaded this work with the support of some members of the Princeton group, Frank included. We opted for a continuous water extraction process, which preserved the composition of the scintillator. We proceeded drawing scintillator from the bottom of the inner vessel, treated via water extraction, and reintroduced from the top. This method, however, had lower efficiency compared to batch processing because the purified scintillator each time is mixed with unprocessed scintillator. Each complete treatment of the total volume resulted in an improvement factor of approximately 2.3. This approach was the only possible one because we certainly could not empty, radio-purify the scintillator, and refill the detector.

We performed four full treatments over a period of about 11 months. The results were excellent: the Uranium content was reduced to just one radioactive nucleus, and Thorium to seven radioactive nuclei, out of 10 billions of billion nuclei of pure material. Radiopurity was greatly enhanced, and other contaminants, such as Krypton, also saw a significant improvement, with a radiopurity factor of about 4.5.

7.5 We Discover What Happens in the Sun and How Its Energy Is Produced

After this second purification, completed in the summer of 2011, we continued collecting data until 2016. This period was referred to as Phase 2 (The data-taking period before the second purification was called Phase One). During this phase, we tackled increasingly complex analyses, focusing on neutrinos from the pep reaction (the fusion of two protons and an electron) and the progenitor pp reaction (the fusion of two hydrogen nuclei). We also revisited the neutrinos from Beryllium −7, achieving a significant reduction in the uncertainty of the final measurement to 2.7%. (In Annexes 7.1, 7.2, and 7.3 you can see how the various neutrino signals and residual contaminant signals still interfere with each other.)

In the meantime, however, we encountered another issue: a leak had developed in the Inner Vessel, causing a small amount of scintillator to mix with the buffer liquid. Within the collaboration, we discussed possible solutions extensively. I was strongly opposed to emptying the detector, repairing the vessel, and restarting all the work, including purification. Such a process would have nullified the results of the two previous purifications. Ultimately, the decision was made not to touch the detector and to monitor the leak closely instead. From that point on, we had to frequently check the shape of the vessel to account for any distortions.

I will now highlight some specific challenges we faced during the study of neutrinos from the pep and pp reactions.

The pep neutrinos exhibit a similar energy profile to those from Beryllium-7, as both are mono-energetic. Their energy signature rises from zero to a maximum when the neutrinos transfer all their energy to the electron they strike. The primary challenge with pep neutrinos, however, is their extremely low rate—only three or four events per day. Moreover, their energy window is dominated by Carbon-11, a radioactive isotope of carbon. This isotope cannot be purified because it is continuously produced by muons (μ particles). As previously mentioned, most muons are blocked by the overburden of the laboratory, but approximately 6 muons/m^2/5 h (per metersquared every 5 hours) still penetrate.

Because Carbon-11 is continuously produced and then is impossible to remove it through purification, its interference must be mitigated using software methods. This is accomplished through two selection processes that reduce its presence to 5%. After these corrections, we could accurately extract the relative neutrino flux (Annex 7.4).

The measurement of the flux of the reaction progenitor of the pp chain, the reaction between two hydrogen nuclei to produce deuterium (this reaction is given the same name as the entire chain, namely pp) has been a remarkable achievement, realized after many years of effort. The challenge in studying this flux lies in the very low energy of its neutrinos, which overlap with the energy range of carbon-14. Additionally, some of these neutrinos have energy levels below the detection threshold, i.e., the minimum energy the detector can detect. While the standard energy threshold is 150 keV, in this case, we managed to reduce it to 120 keV (Annexes 7.2 and 7.3).

We also measured the fusion reaction involving Boron-8. Its flow is lower compared to all the other reactions we studied, but it has the highest energy, reaching up to 15 MeV. It was possible to study it, but only starting at energies above 3.2 MeV due to the presence of a residue of Thallium-208 a contaminant that emits radiation up to about 3 MeV.

Finally we measured all the fusion reactions emitting neutrinos of the proton-proton chain, which generates 99% of the Sun's energy. *We thus completed a journey that began in the 1930s when Hans Bethe proposed that solar energy was produced via the proton-proton chain.*

By comparing the fluxes we measured with the predictions of the Standard Solar Model, we found good agreement within the uncertainties of both the experimental measurements and the model's predictions. This result generated immense interest within the astrophysics community because, for more than 80 years, no one had succeeded in proving the existence of the individual fusion reactions in the Sun that emit neutrinos.

This discovery, summarized in an article published in *Nature*, one of the most prestigious scientific journals, in 2014, was recognized by Physics World of the British Institute of Physics (IOP) one of the ten best scientific achievements worldwide. Additionally, the Italian Post Office issued a commemorative stamp to mark the accomplishment. Personally, I was honored to receive the international Bruno Pontecorvo Prize, awarded by a committee of Canadian, Japanese, Italian, and Russian physicists, as well as the Enrico Fermi Prize, the highest Italian award in physics, while the head of the Polish

Fig. 7.4 Awards and aknowledgements received by Borexino, and myself in 2014 (Pontecorvo price) and 2017 (Fermi price), and by Marcin Wójcik in 2009

group from Krakow, Marcin Wójcik, received an award from the Polish Prime Minister (Fig. 7.4).

The reader might ask: is it really so important to understand the mechanism that makes the Sun shine? There are many answers to this question.

The first, very general answer, is that the pursuit of knowledge is a continuous challenge for humanity. As a late friend of mine, a biologist, always said: "Man's shirt is too tight for him." By this, he meant that humans always feel a sense of incompleteness and are driven to push beyond their limits.

The second, more specific answer is that the study of the heavens above us, carried forward through the work of many scientists, had a missing piece: how is the Sun so powerful that it can provide light and heat to the entire solar system? As explained earlier, the discoveries made in this experiment about the Sun, which will eventually extend to stars, provide answers to the questions of all those who, over millennia, gazed at the lights in the sky and wondered why they shone. I believe that for all of us, it is not uninteresting to understand how everything around us is made and how it works.

The third answer concerns science as a whole. Science, as one of the foundations of knowledge, precedes technology; the latter partially could not exist without the understanding of phenomena gained through fundamental science. Time and again, discoveries thought to be of no practical use beyond advancing human knowledge have proven otherwise. There are countless examples to illustrate this point.

But there is also other perspectives. While we were studying solar neutrinos, a British journalist came to Gran Sasso to interview me. He later sent me a draft of his article for approval. In the closing lines, he described his return

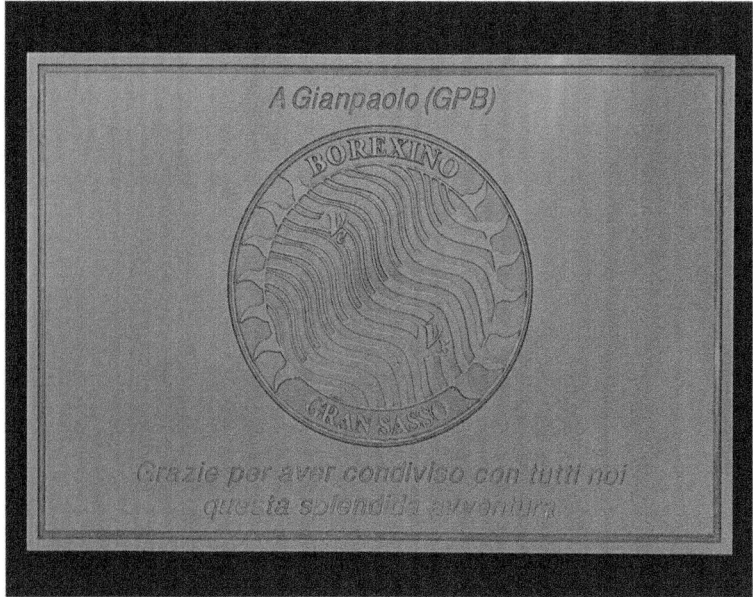

Fig. 7.5 Plaque prepared by researchers and technicians of Borexino—It is dedicated: to Gianpaolo (GBP) and the engraved phrase reads: "Thank you for sharing this splendid adventure with all of us"

to Rome as the Sun was setting, painting the sky red. He wondered whether understanding how the Sun works and knowing it emits an immense quantity of neutrinos somehow diminished the mystery and poetry surrounding our star. Similarly, he mused about whether the moonlit nights, with our knowledge—after the astronauts' journeys—of what the Moon is made of, had lessened the magical influence it once had on lovers' emotions.

On a personal note, I would like to add that this success not only marked a significant achievement but also relieved me of the worry I had for the young physicists working on this experiment. Had we failed, they would certainly have suffered setbacks in their scientific careers. I was also deeply gratified to receive expressions of gratitude from the physicists and engineers of Gran Sasso, members of Borexino, who thanked me for involving them in this extraordinary adventure (Fig. 7.5).

Annex 7.1

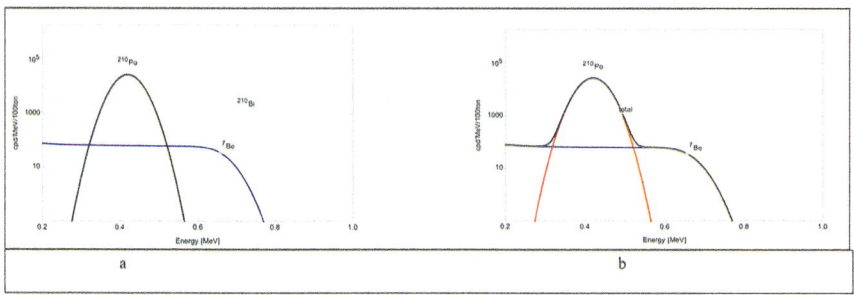

a

b

In these plots, the energy distributions can be observed. On the horizontal axis, the energy values are displayed, while on the vertical axis, the frequency is shown, indicating how often a specific energy value corresponds to that of neutrinos or, in case of ^{210}Po (Polonium-210), alpha particles, which are simply Helium nuclei.

In Figure a, the energy distribution of neutrinos produced by the reaction with Beryllium-7 is represented by the blue curve, while the shape of the alpha particles (Helium nuclei) emitted by Polonium-210 is shown in green. Figure b displays the actual observed data: in the energy range of 300–600 keV, the black line represents the sum of the blue line (beryllium neutrinos) and the red line (polonium emissions).

The goal of the analysis is to disentangle the Beryllium neutrinos from the overall data distribution. This task is further complicated by additional contributions, as highlighted in Annex 7.2, where the actual plot is shown. Despite these challenges, this has been one of the simplest analyses performed in the Borexino experiment.

Annex 7.2

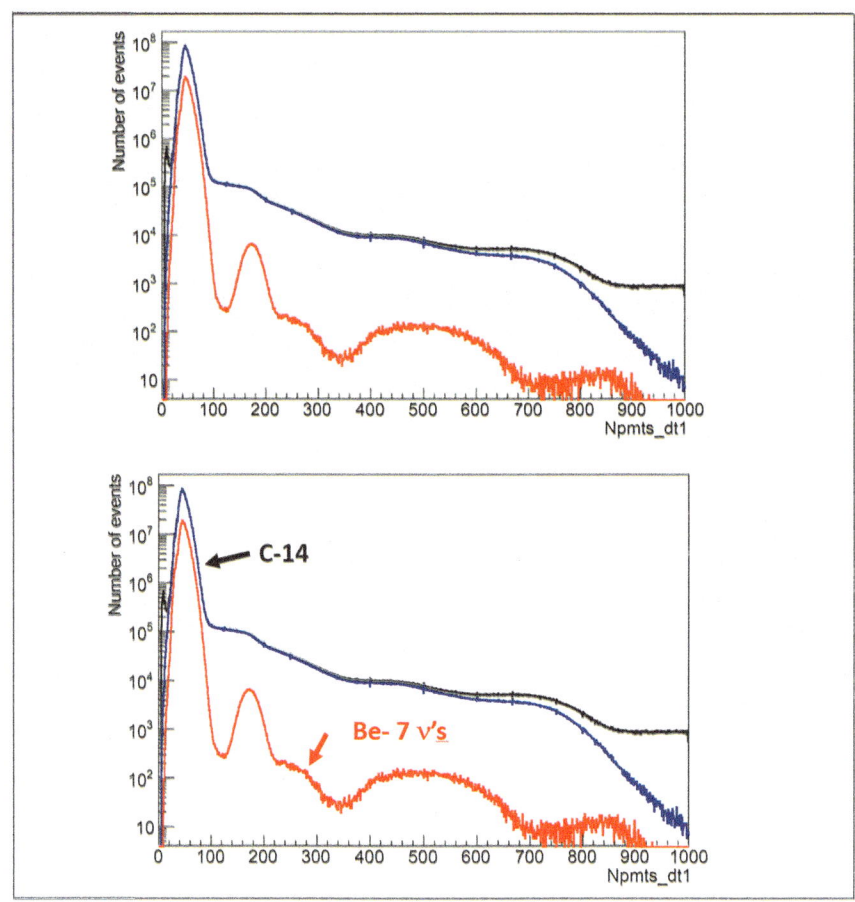

In the first figure, all the received and reconstructed data are represented by the black line. From this, signals arriving from the photomultipliers located in the water tank—tracking the passage of a muon particle unrelated to the neutrinos under study—are subtracted. In addition to this subtraction, all the fake signals originating from external sources are removed, leaving the red line as the final result.

In the second figure, the same energy distributions as in the first are shown, with certain peaks or humps indicated. In the red part, the energies corresponding to Polonium-210 and Beryllium-7 are observed. These align with figure b of Annex 7.1, but the differences are attributed to the presence of signals from other contaminants, including Carbon-14. Therefore, the neutrinos from Beryllium-7 must be extracted solely from the hump marked in red with the arrow labeled "7Be".

Annex 7.3

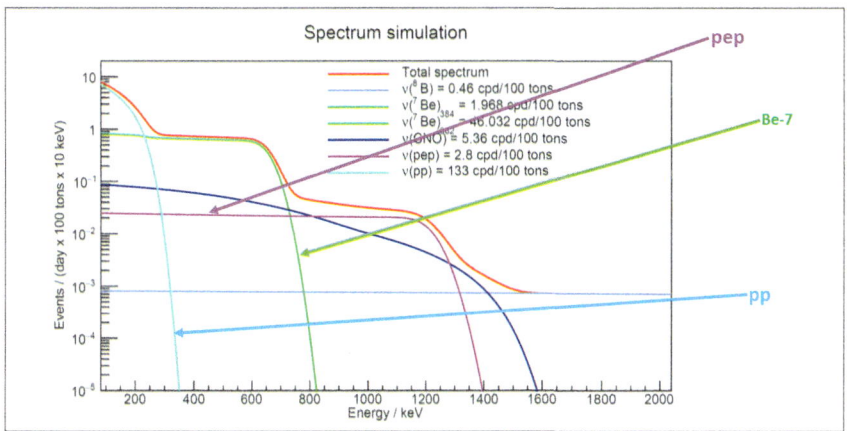

The red line represents the total spectrum; the sky blue, green, and violet lines correspond, respectively, to the energy spectra of neutrinos from pp., Beryllium-7, and pep. The blue line will be discussed in the following chapters.

Annex 7.4

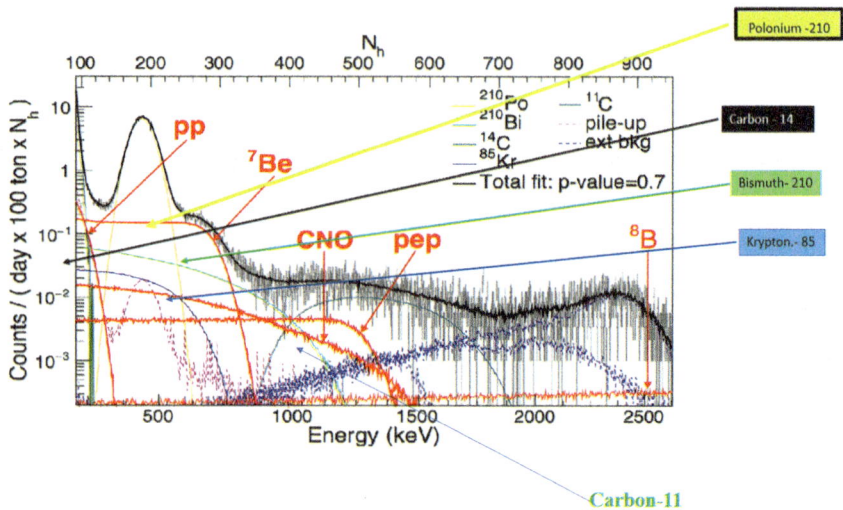

A complete plot with the neutrino signals from solar reactions and the residuals of contaminants. This is the plot that the analysis work managed to reconstruct using a substantially complex algorithm, which succeeded in building everything from the general shape represented here by the black line and by knowing the shape of the neutrino signals corresponding to all possible fusion reactions and of the contaminants. This is the result of many years of work—about 7 years of data acquisition, reconstruction to determine the energy and the spatial point where the interaction occurred, and finally, interpretation (inserted in papers: Nature, vol. 562, 25 October 2018, 505, and in Annu. Rev. Nucl. Part. Sci. 74:369–88, 2024).

8

Sun Stability and Earth Orbit

8.1 The Sun Stability

The luminosity of the Sun, which is the amount of solar light energy reaching the Earth per unit of time, is typically determined by measuring the photons that arrive at the Earth. This allows for a comparison between the two methods of measuring solar luminosity: via photons and via neutrinos, remembering that the fusion reactions of the pp chain produce 99% of all solar energy. From the neutrino fluxes, we can deduce the fusion rate occurring in the Sun and, consequently, the rate of the pp chain, each of which releases 26.73 MeV of energy. This energy is transformed into energy photons which travel outward through a process of diffusion, absorption, and thermal conversion.

From the comparison between the luminosities measured via photons and via neutrinos, we find that the two measurements are in agreement, indicating that the Sun has been stable over a timescale of 100,000 years. Neutrinos take only a few seconds to exit the Sun and about 8 min to reach the Earth, while photons take an average of 100,000 years because they follow a *random walk* (Fig. 8.1), being deflected in all directions, and undergoing repeated absorption and re-emission processes. This means that the sunlight shining on us now was produced on the average 100,000 years ago.

While one might note that a time of 100,000 years is negligible compared to the lifespan of a star, which exceeds billions of years, it is remarkable that we can directly confirm this stability.

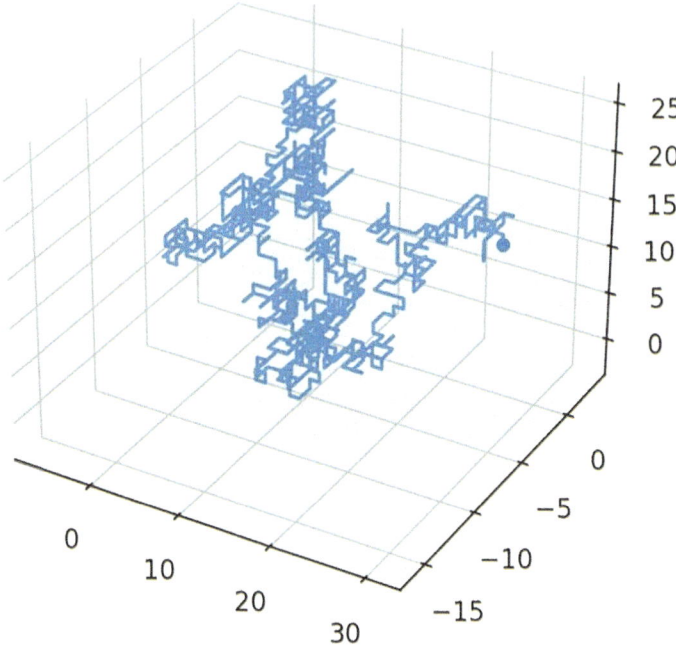

Fig. 8.1 A random walk

8.2 Calculating the Earth's Orbit Using Solar Neutrinos

The shape of the neutrino energy distributions, which align with those expected from the nuclear reactions of the four fusion processes that emit neutrinos, provides evidence that the neutrinos originate from the Sun. However, there is a more direct way to demonstrate that the observed neutrinos are of solar origin: by taking advantage of the eccentricity of the Earth's orbit, which is an ellipse with the Sun at one of its foci. Naturally, when the Earth is closer to the Sun, the flux of neutrinos captured by the Earth is greater than when it is farther away. As a result, over the course of a solar year, we observe a solar neutrino flux that varies, reaching a maximum when the distance is smallest (in winter) and a minimum when the distance is greatest (in summer) (Fig. 8.2).

8 Sun Stability and Earth Orbit

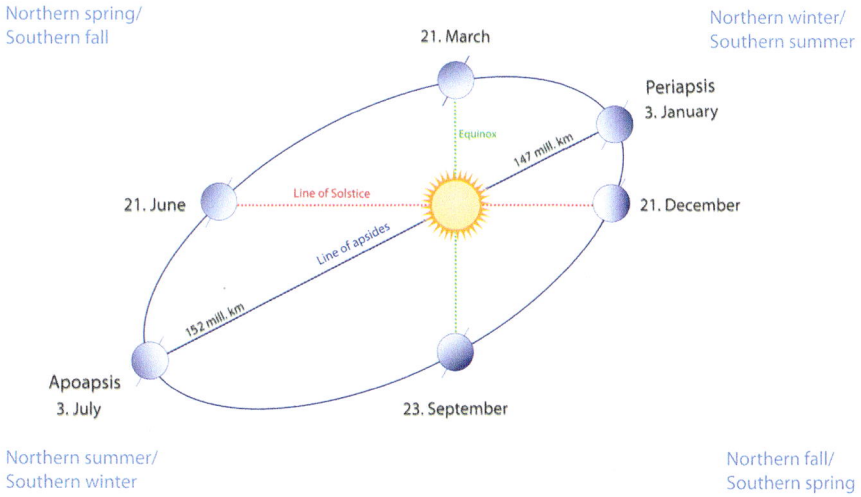

Fig. 8.2 Earth's orbit. (*Source: By Isadora Ibiza—File:Earth poster5.svg-Wikipedia*)

This result is in excellent agreement with theoretical predictions of the phenomenon. We have thus directly demonstrated the solar origin of the observed neutrinos.

Furthermore, measuring this modulation of the solar neutrino flux has enabled the reconstruction of the Earth's orbit by determining the Earth-Sun distance on a month-by-month basis, resulting in the most accurate reconstruction of the Earth's orbit ever achieved using neutrinos (see also Annex 8.1).

Annex 8.1

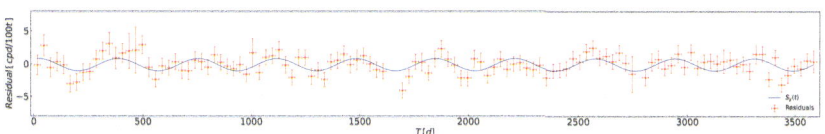

Variation of the neutrinos solar flux with respect to the average of the solar neutrino flux collected over 10 years. The horizontal axis shows the time in month unit while the vertical axis shows the month-by-month difference of the measurement with respect to the average of the flux. The experimental data are in red with the uncertainty of the measurement represented by the vertical bars; the blue line is the result of the best fit, that is, the search for the best shape which represents the experimental data, obtained with the computer through an algorithm. The maxima correspond to the neutrino flux when the Earth is at periapsis, that is, on January 3, while the minima relate to the neutrino flux when the Earth is at apoapsis, on July 3 (see Fig. 8.1).

Correspondence between the neutrino flux and the Earth's position.

9

And the Stars?

9.1 A Fourth Neutrino?

One member of the collaboration, Marco Pallavicini, was advocating for measuring the possible, though unlikely, fourth neutrino using the Borexino detector. This project would involve a large radioactive neutrino source placed just below the detector. I was not very convinced by this idea, as it would have made the measurement of CNO impossible.

A fourth neutrino had been hypothesized by symmetry with other particles that share some similar characteristics. It was called "sterile" because it did not interact. At that time, there was already a Russian experiment that had found no evidence for the existence of this neutrino.

Marco Pallavicini's proposal was to use a very powerful source that would emit almost exclusively neutrinos. The other radiations emitted by the source could be absorbed by a casing immersed in a calorimeter, which would measure the heat emitted by the radioactive decays of the source. Marco Pallavicini managed to obtain financial support from the European Union and then established a collaboration with a French group from Saclay and with a Russian group that would supply the source, enriched in a power reactor. Needless to say, we were criticized by environmentalists, who argued that the source would be very dangerous and demanded that the project be prohibited.

Despite the importance of research of this kind, I wasn't particularly interested in this measurement and had tried to delay it in favor of the CNO cycle measurement. However, I found little support in the collaboration, except from a few people who disagreed with the project, such as Frank Calaprice, who stated he would not participate in the measurement, and in particular a young researcher from Gran Sasso, Nicola Rossi. Nicola continued to work on the CNO analysis with the data already taken, despite criticism from other analysis group members who were convinced, as much of the collaboration was, that we would never be able to measure CNO. I remember receiving an email from a German collaborator who accused me of wasting people's time by insisting on an unfeasible measurement. My response was that I considered it possible and reminded him that the whole Borexino project had been filled with skepticism, which had always been proven wrong by the facts.

The Pallavicini's project was progressing well for the Italian and French groups, and both teams were building a calorimeter. One of these was tested in the pit beneath the Borexino detector, which was the position it would occupy during the planned measurement. However, the Russian group, which was responsible for preparing the high-intensity source, declared after several months and a few attempts that the technique they had available had not allowed to obtain the source. As a result, the fourth neutrino search project with Borexino was abandoned.

At this point, the project to search for the CNO cycle was resumed, and even some members of the collaboration, who had been skeptical about the feasibility of such a measurement or had even declared it impossible, began to work on the project, driven by my constant encouragement.

9.2 How Do We Investigate Stars?

The temperature that the pp chain reaches in the core of the Sun is about 15 million degrees. This temperature generates pressure that counterbalances the gravitational force which push solar masses against each other, thereby preventing the Sun's implosion. The temperature is directly linked to the chaotic motion of particles, and this motion increases with rising temperature.

This process occurs in stars of similar or smaller dimensions than the Sun. However, for massive stars, those with at least 30% greater mass than the Sun, the temperature reached by the pp chain is insufficient to prevent implosion because the gravitational attraction between masses increases with greater mass.

In the 1930s, Hans Bethe, along with Carl F. von Weizsäcker, hypothesized that in massive stars, a nuclear fusion cycle dominates, still based on hydrogen burning but catalyzed by carbon, oxygen, and nitrogen—thus called the CNO cycle (catalysts do not take part in the reaction but influence it by facilitating, accelerating, and enhancing it). This cycle produces temperatures approximately ten times higher than those of the pp chain: about 150 million degrees. This temperature is sufficient to counteract the gravitational force in massive stars and prevent their implosion.

But how can this cycle be definitively proven to exist? Observing neutrinos from distant stars is unfeasible, as Earth's immense distance reduces our planet to an infinitesimal point from the perspective of stellar neutrinos. Only in the exceptional case of a supernova explosion, neutrinos from stars can be observed on Earth. These explosions release enormous energy, resulting in a burst of neutrinos emitted over a very short time. If the supernova's distance is not too great, a flood of neutrinos arrives in a few seconds on Earth.

The standard solar model predicts that 1% of the Sun's energy comes from the CNO cycle, while 99% is produced by the pp chain. Our task was to measure the neutrinos emitted by the CNO cycle in the Sun. Studying the CNO cycle has been a significant effort lasting 4 years. The challenge stemmed from the extremely low flux of neutrinos produced by this cycle in the Sun, compounded by interference from residual radioactive isotopes in the scintillator. In the energy range where CNO neutrinos are expected, three overlapping distributions complicate detection: signals from the pep reaction, the CNO cycle, and the isotope Bismuth-210.

The pep signal posed no issues, as we had already measured it, and the Standard Solar Model's predictions for it are highly reliable. However, the challenge arose with CNO and Bismuth-210 because their energy distributions are very similar. Attempting to separate their contributions using a computer algorithm returned only their combined total. Therefore, it was necessary to independently measure the rate of interactions due to Bismuth-210 neutrinos. This was achieved thanks to the presence in the scintillator of residues of another isotope, Polonium-210 (to understand better, see Annex 9.1).

Within the natural radioactivity family headed by Uranium-238, these two isotopes are produced by Lead-210. In this family, a secular equilibrium is established, ensuring all isotopes have the same activity: a daughter isotope can decay only after the decay of its parent isotope. Thus, Polonium-210, a daughter of Bismuth-210, shares the same activity and decay frequency as its parent (Fig. 9.1).

To determine the activity of Bismuth-210, it was therefore enough to measure the emissions of Polonium-210. However, two complications arose. First, also Polonium not in secular equilibrium was present in the scintillator, introduced during filling and collected along transport lines. Second, convective motions inside the scintillator, caused by temperature inhomogeneities, carried Lead-210 residues (which produce Bismuth-210 and then Polonium-210), from particulate matter and residual dust on the Inner Vessel's internal surface, into the scintillator.

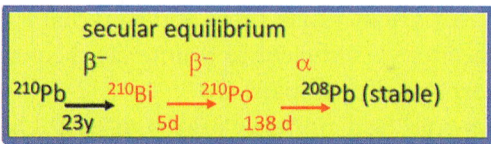

Fig. 9.1 This sequence is part of the family of natural radioactivity whose progenitor is Uranium-238. All these nuclides, being in secular equilibrium, exhibit the same rate of decay, i.e., their transformation into the subsequent nuclide, emitting radiations. In this figure, above the nuclides, the type of radiation they emit during decay are indicated: β^- represents a negative electron, while α represents a Helium nucleus. Below the nuclides, the average times required for their decay, i.e., their transformation into the next nuclide, are indicated. Diagram illustrates the decay chain of Lead-210 in secular equilibrium. It shows Lead-210 decaying to Bismuth-210 through beta decay over 23 years, then to Polonium-210 through beta decay over 5 days, and finally to stable Lead-208 through alpha decay over 138 days

For Polonium not in secular equilibrium, it was enough to wait sufficient time for it to decay into Pb-208, which is stable. This way, only Po-210, continuously produced in the sequence starting from Pb-210, remained, thus exhibiting a constant rate.

To suppress convective motions, it was necessary to stabilize the temperature of the detector, which was a challenging problem because the detector was placed in a large room that received air from outside, making precise stabilization impossible. We therefore decided to thermally insulate the entire detector using a thermal insulator (Fig. 9.2). After applying the insulation, we had to wait some time for the temperature to stabilize, which we then monitored using multiple temperature probes installed inside and outside the water tank (Fig. 9.3). The operation was quite complex because the entire large tank, including the organ pipes, had to be covered with insulation material. However, the effort was rewarded with a stabilization level of 0.07°.

Finally, we succeeded in measuring the rate of Polonium-210 events and, therefore, due to the secular equilibrium, the rate of Bismuth-210. Knowing the rate of this latter, it was relatively straightforward to extract the CNO signal with high reliability using computer algorithms.

Borexino and all of us had won the challenge; we had demonstrated, for the first time, the existence of the CNO cycle, which dominates in massive stars.

Shortly after the start of the CNO measurement, the coordination of the analysis alternated between Gemma Testera and Barbara Caccianiga with Livia Ludhova; the coordination of Barbara and Livia was very important for achieving the objective of the CNO measurement.

The analysis of the CNO cycle was primarily conducted by Italian groups, alongside a team quickly assembled by Livia Ludhova, who had relocated to

9 And the Stars? 133

Fig. 9.2 The Borexino external tank is encased in insulating material

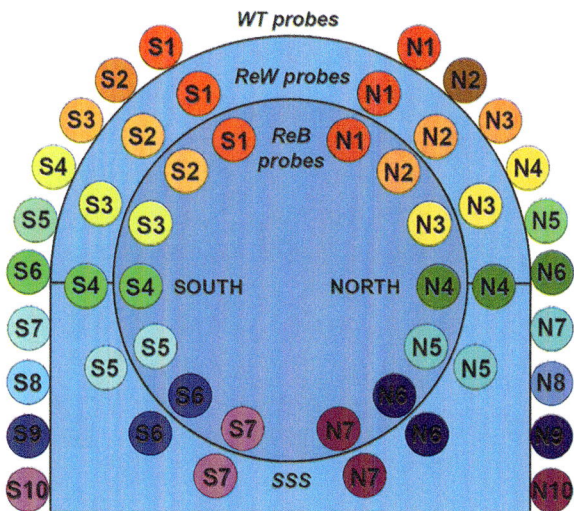

Fig. 9.3 Placement of the various temperature probes installed within the water tank. The left colored dots indicate temperature gradients

Jülich, Germany. This effort also saw contributions from physicists in the Krakow and Dubna groups, as well as PhD students from the Gran Sasso Science Institute tutored by Francesco Vissani, a staunch supporter of Borexino. Additionally, members of the Munich group, having moved to other institutions, continued to contribute to the CNO analysis. Among these, the contribution of J. Martyn from the University of Mainz is noteworthy, as he developed a method to obtain information on the directionality of incident neutrinos, even in a scintillator like Borexino's. This method proved to be very useful in the analysis of the CNO cycle.

During this time new institutions joined our effort, offering financial support for researchers no longer funded by INFN or by the Gran Sasso Laboratory, such as George Korga and Laszlo Papp. As a result, the number of institutes and universities participating in the collaboration grew significantly (Figs. 9.4 and 9.5).

While INFN had continuously supported us, we recognized that the daily on-site costs for maintaining all the equipment, including Borexino's ancillary systems, were substantial. To address this challenge, we decided to establish Common Funds—general contributions to cover the experiment's expenses. Each institute contributed an amount typically determined at its discretion. These funds enabled us to manage ongoing expenses, which were significant,

Fig. 9.4 The composition of the Borexino Collaboration during the period of the CNO analysis

Fig. 9.5 The Borexino Collaboration in the second decade of the 2000s. The structure in the background is the support for the CTF photomultipliers, which was preserved by the laboratory and installed outdoors

and to support operators whom we otherwise would have been unable to retain.

The CNO measurement garnered significant attention, as it had been anticipated for 90 years. We published the findings in *Nature* in 2020, which featured the result on the cover of the issue containing the article. Again Physics World, of the British Institute of Physics (IOP), recognized this discovery as one of the top ten scientific breakthroughs of 2020 (Fig. 9.6).

9.3 The Sun Metallicity

The data obtained by Borexino on the Sun have provided further insights into its structure, particularly concerning its metallicity.

Metallicity refers to the presence of elements in the Sun with an atomic number higher than hydrogen, which has the lowest atomic number in the periodic table of chemical elements. It is expressed by the ratio Z/X, where Z represents elements with an atomic number greater than 2, and X represents hydrogen. The Sun's metallicity is determined through spectroscopic measurements of its photosphere. These measurements identify the light emitted

Fig. 9.6 In 2020, the results of Borexino, which discovered experimentally the existence of the CNO cycle dominating in massive stars, were ranked among the world's top 10 by *Physics World* of the British IOP

by atoms, the wavelengths of which serve as unique "fingerprints" for each chemical element.

To estimate the metallicity of the solar atmosphere, scientists extrapolate photosphere measurements using various models. Some models yield a Z/X ratio of 0.023, referred to as *High Metallicity*, while others give a value of 0.020, termed *Low Metallicity*.

Two main methods are used to assess which of the two metallicity values is more realistic: helioseismology and solar neutrino flux measurements (I recall what was mentioned earlier: helioseismology is the study of the motions occurring on the Sun's surface caused by the propagation of pressure waves). Helioseismology shows a much stronger alignment with high-metallicity models. Regarding neutrino fluxes, the Standard Solar Model predicts significantly different values for key fluxes, such as those of Boron-8 and the CNO cycle, depending on the assumed metallicity. Although the evidence strongly supports high metallicity, it remains not higly conclusive. Nevertheless, the data consistently indicate a preference for the high-metallicity scenario.

The debate over solar metallicity has been a major challenge for the Standard Solar Model for more than 50 years. Borexino's contributions have been crucial in improving our understanding and bringing us closer to solving this long-standing puzzle.

9.4 Annex 9.1

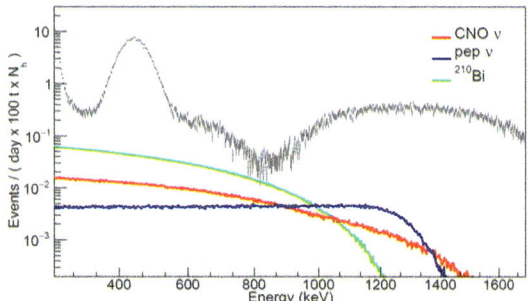

The plot shows the neutrinos interactions rates depending of energy in keV from the pep reaction, (blue) CNO cycle (red) and Bismuth-210 (green.). The jagged black line corresponds to all the Borexino data as they appear when they are collected and processed. The y-axis represents events per day per 100 tons, and the x-axis represents energy in keV.

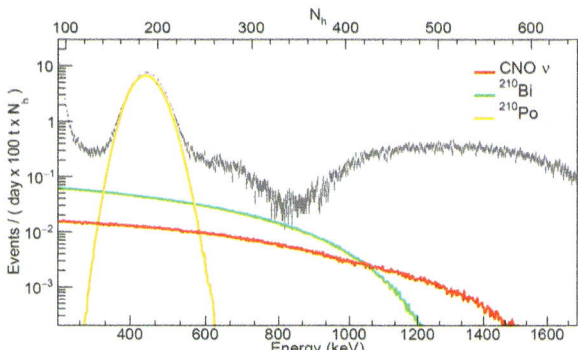

Same plot with the inclusion of Polonium-210. As explained in the text, the energy shape and position corresponding to neutrinos from CNO and Bismuth-210 do not allow for the extraction of CNO data through software alone. However, measuring the contribution of Polonium-210, which is in secular equilibrium with Bismuth-210, enables an independent calculation of the Bismuth-210 contribution, thereby allowing the extraction of the CNO component.

10

Corollaries

10.1 Neutrino Oscillation

As explained earlier in Chap. 3 regarding neutrinos, they exhibit the phenomenon of oscillation. During their journey, a neutrino can change its "identity," transforming into one of its two siblings. For instance, a solar neutrino, initially an electron-neutrino, can transform into a muon-neutrino or a tau-neutrino during its 8-min journey from the Sun to the Earth.

This oscillation depends on the energy of the neutrino, and thus the survival probability of the electron-neutrino—the probability that it remains unchanged—varies accordingly.

Before Borexino, these probabilities were only approximately known because no experiment had successfully measured the fluxes of the various nuclear fusion reactions occurring in the Sun individually, except for the oscillation in matter measured through the reaction involving Boron-8 by Superkamiokande with good precision.

Borexino successfully measured the fluxes of different fusion reactions over a wide range of energies, from oscillation in vacuum at low energy to oscillation in matter at higher energy. Vacuum oscillation describes how neutrinos change as they travel from the Sun to the Earth in the absence of matter, while matter oscillation occurs when neutrinos pass through regions influenced by matter. At very low energies, the effect of matter becomes negligible.

This accomplishment is particularly significant because Borexino was the first experiment to measure the electron-neutrino survival probability in both the vacuum and matter. Measuring both regimes with the same experiment ensures consistency and eliminates potential issues related to detector discrepancies. This unified approach represents the first time a single experiment has conducted such a comprehensive and reliable measurement.

10.2 Geoneutrinos

This final section diverges slightly from the primary focus of this book, which has been on the mechanisms that make the Sun and stars shine. However, it shares a common theme: the unique ability of neutrinos to serve as formidable probes, providing insights into otherwise inaccessible regions. In this case, the focus shifts to the Earth's interior.

The use of elementary particles for geophysical purposes is advancing rapidly. For example, muons—particles we discussed in earlier chapters—are now employed to study volcanic activity and assess the likelihood of eruptions. While the Sun and stars have been our primary focus, here we turn to the study of our planet in the solar system: the Earth. Specifically, we explore geoneutrinos, which are antineutrinos—the antiparticles of neutrinos—originating from the Earth's interior. These particles are produced by the decay of radioactive isotopes present in the Earth's crust and mantle (Fig. 10.1).

While radioactive isotopes in the Earth's crust can be directly studied by analyzing surface sediments and rocks, this approach is not feasible for the mantle, which remains inaccessible for direct sampling. Geologists traditionally investigate the Earth's interior by analyzing the propagation of seismic waves—both longitudinal and radial—which reveal whether the Earth crossed materials are solid or liquid. However, the chemical composition of the Earth's interior remains largely unknown, except for insights gained from volcanic processes, tectonic activity, and samples from ocean ridges, where mantle rocks occasionally surface.

The mantle is also known to exhibit temperature heterogeneity, partly due to the uneven distribution of radioactive decays. This creates convective currents between hotter and cooler regions, driving plate tectonics. These plate movements can lead to collisions, causing earthquakes.

The study of geoneutrinos can provide valuable information about the presence of radioactive elements in the mantle and their radiogenic contribution to the Earth's heat. As mentioned earlier, the radioactivity of the Earth's crust is directly measurable and, therefore, well understood. The radioactive

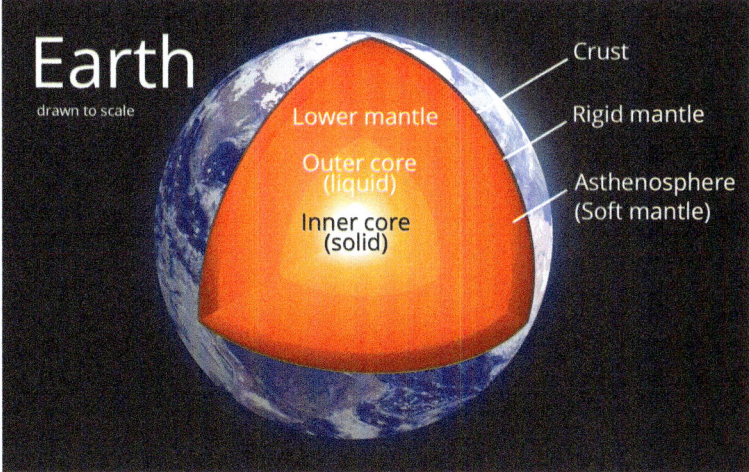

Fig. 10.1 Schematic cross-section of the Earth. The Earth has an onion-like structure, with a radius of 6378 km at the equator. Its central core includes an inner solid portion (radius 1220 km) and an outer liquid portion (2895 km thick). Surrounding the core is the viscous mantle (about 2000 km deep), on which tectonic plates float and move. There are two types of crust: the oceanic crust (approximately 10 km thick) and the continental crust (varying from 20 to 70 km thick). (Source: By Isadora Ibiza—File:Earth poster.svg—Wikipedia)

isotopes within the Earth belong to three families of natural radioactivity, originating from Uranium-238, Thorium-232, and Potassium-40.

The flux of geoneutrinos is much lower than that of solar neutrinos. In Borexino, only two geoneutrinos are observed every 3 months. These particles can be identified through a reaction induced by antineutrinos in the scintillator, which produces signals with specific characteristics that make them easily recognizable.

This detection reaction can only occur above a certain energy, which excludes the detection of antineutrinos produced by the decay of Potassium-40, as well as a portion of those emitted from Uranium-238 and Thorium-232. The study of geoneutrinos is further complicated by internal contaminants, due to contamination in the scintillator, and antineutrinos originating from nuclear reactors. In Borexino, internal radioactivity is negligible, and the flux of antineutrinos from reactors is present but limited, owing to the relatively small number of reactors located far from Gran Sasso, primarily in France, Switzerland, and other European countries.

To estimate the total reactor antineutrino flux, calculations are performed for all 440 reactors worldwide. These calculations take into account the type of fuel used, variations in reactor activity during the data collection period, and other relevant parameters. The necessary data are obtained from the databases of the International Atomic Energy Agency (IAEA) and Electricité de France.

Geoneutrinos have higher energies than solar neutrinos, though their energy ranges overlap to some extent. As a result, the radiopurity requirements for geoneutrino detectors are less stringent than those for solar neutrino experiments. While Borexino is uniquely equipped to study solar neutrinos, it is not the only experiment capable of measuring geoneutrinos. The Japanese KamLAND experiment, with a detector volume three times larger than Borexino's, can also detect geoneutrinos.

Although the energy distributions of reactor antineutrinos and geoneutrinos overlap only partially, separating the contributions from different sources is essential. To evaluate the flux of geoneutrinos originating from the Earth's mantle and estimate the relative density of radioactive nuclides, we first need to subtract the contribution from the crust. This requires distinguishing between geoneutrinos from the continental crust (where Borexino is located) and those from the oceanic crust (which partly contributes to KamLAND).

This operation is far from simple because, first of all, it requires a good assessment of the radioactivity of the crust, which must be extracted from the number of geoneutrinos that can be measured in a location. This assessment must take into account not only the emissions near the measurement site but also those from distant sources. Once the subtraction is performed between the total number of measured geoneutrinos and those related to the crust, it is necessary to evaluate what this number means in relation to the radioactivity of the mantle. This is equally challenging because this evaluation differs depending on whether we consider the radioactive nuclides to be distributed homogeneously within the mantle or concentrated at its base, at the boundary with the Earth's core.

In Borexino, data collection for geoneutrinos took place from December 2007 to April 2019, covering a total of 3263 days. After all the necessary selections, the total number of geoneutrinos interactions obtained was 154. Without delving into model-based assessments, I can report here the result concerning the percentage of Earth's heat attributed to radioactive elements within the planet.

It's worth recalling that Earth's heat refers to the heat that the planet continuously loses and radiates into the surrounding space. Borexino's estimate for the radiogenic heat his quite high, amounting to 38 Terawatts (TW), equivalent to 38 billion Watts. This is to be compared with the total Earth heat, which is estimated to be between 44 and 47 TW. As we can see, this represents a very high percentage. If Borexino's result is combined with that of the Japanese KamLAND experiment a lower value of 21 TW is obtained.

To achieve more reliable values for Earth's internal radioactivity, it is necessary to collect many more signals from solar meutrinos interactions, which means taking data over many years, as the number of neutrinos coming from within the Earth is not particularly high.

To better understand what Borexino achieved by detecting geoneutrinos, one can refer to the plot in the Annex 10.1.

Annex 10.1

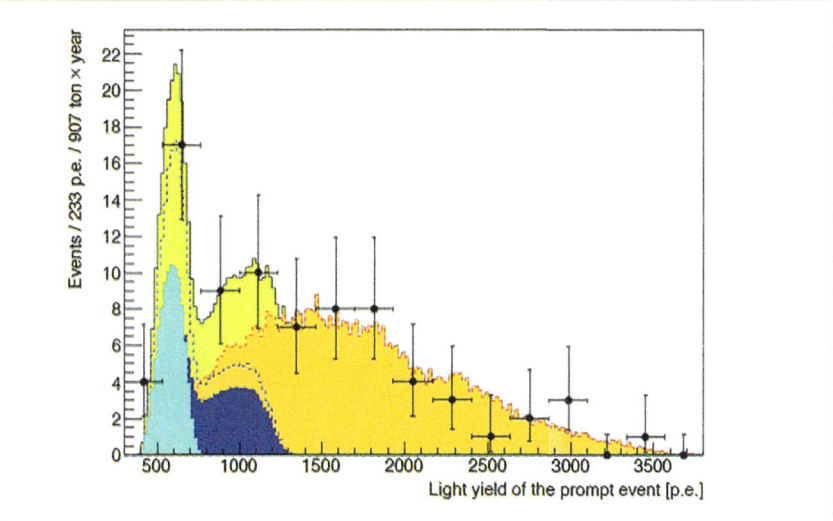

Geoneutrino Energy Distribution in Borexino.
The horizontal axis represents the energy, while the vertical axis shows the frequency of interactions having the energy indicated in the horizontal axis. The black dots represent all the data obtained, with vertical bars indicating the statistical uncertainties. The ochre-colored distribution corresponds to reactor antineutrinos, while the yellow distribution represents geoneutrinos. The geoneutrino distribution is further divided into the blue section, representing antineutrinos emitted during the decay of Uranium-238 family components, and the light blue section, representing antineutrinos produced from Thorium-232 decay.

Geoneutrino Energy Distribution in Borexino

The horizontal axis represents the energy, while the vertical axis shows the frequency of interactions having the energy indicated in the horizontal axis. The black dots represent all the data obtained, with vertical bars indicating the statistical uncertainties. The ochre-colored distribution corresponds to reactor antineutrinos, while the yellow distribution represents geoneutrinos. The geoneutrino distribution is further divided into the blue section, representing antineutrinos emitted during the decay of Uranium-238 family components, and the light blue section, representing antineutrinos produced from Thorium-232 decay.

11

Farewell Borexino

11.1 Farewell Borexino

Borexino was decommissioned in October 2021. While we could have continued for a few more months to gather additional data and further reduce the uncertainty in the CNO measurements, the decision to shut down was ultimately driven by mounting pressure from environmental activists. Nonetheless, we successfully achieved the much more than program's primary objectives.

Data analysis continued for few years after the shutdown, and even now, a final analysis is nearing completion.

Even after its closure, Borexino's groundbreaking results have remained a focal point of interest, as evidenced by the numerous invitations to present them at international conferences. In 2021, the European Physical Society recognized the importance of our work by awarding the Borexino collaboration with the biennial Giuseppe and Vanna Cocconi Prize. Additionally, Frank Calaprice in 2023 received the Bethe Prize from the American Physical Society, the Polish Borexino group was honored in 2022 with an award from the Polish Ministry of Science and Higher Education, Livia Ludhova in 2025 has been awarded of the Order of Ľudovít Štúr, 2nd Class, Civil Division by the President of the Slovak Republic (Fig. 11.1).

On behalf of the Borexino collaboration, I authored a comprehensive review of the experiment's achievements in a highly selective journal and had the privilege of delivering the first talk at the Neutrino 2024 world conference, again on the Borexino achievements.

Thus ends an adventure that lasted half a lifetime.

Fig. 11.1 Further awards received by the Borexino collaboration and by physicists who have been members of Borexino, Frank Calaprice, Marcin Wójcik, and Livia Ludhova. On the left, a gold medal from the European Physical Society labeled "G. & V. Cocconi Prize 2021 - EPS." (Borexino collaboration), next, the APS logo with "Bethe Prize 2023" (Frank Calaprice). To the right, the logo of the Polish Ministry of Education and Science with "Award of Polish Science Minister 2022." (Polish Borexino group)) On the far right, a medal of Order of Ľudovít Štúr, 2nd Class, Civil Division by the President of the Slovak Republic (Livia Ludhova)

11.2 The Borexino Legacy

Borexino was not only a highly successful experiment but also a remarkable human adventure. Over the 31 years of the experiment, many people contributed—physicists, engineers, and technicians—some working on it for years, while others made occasional contributions. The number of physicists, engineers, and technicians simultaneously involved in Borexino never exceeded 100–120 individuals at any given time. Those who started working with me in 1990 and continued until 2021 are fewer than 15. Sadly, over these 31 years, 14 collaborators passed away. On the other hand, several children were born to young couples, where one or both parents were involved in the experiment, adding a human polish Borexino group awards touch to this extraordinary scientific journey. A particularly positive aspect of the experiment was the involvement of numerous young people, who joined progressively or replaced others over time, including in the last years.

The legacy of Borexino is multifaceted. First and foremost, the techniques developed by Borexino to achieve an unprecedented level of radiopurity will undoubtedly be useful for experiments that aim to study low-energy neutrinos, either now or in the near future. This technology is already being utilized by other ongoing experiments requiring low radioactivity, including Juno in China, as well as experiments focused on dark matter and neutrino-less double beta decay. This is certainly a legacy Borexino leaves for future experimental efforts.

There is also a scientific legacy, stemming from the discoveries we have made related to the Sun, stars, neutrino physics—especially the oscillation phenomenon—and geoneutrinos. Borexino's discoveries encompass stars of all sizes: stars similar to or smaller than the Sun are powered by the proton-proton (pp) chain, while massive stars, at least 30% larger than the Sun, rely on the CNO cycle.

As for Borexino's measurements of geoneutrinos, I believe they will be surpassed by larger experiments such as SNO+ and Juno. However, for solar neutrinos, achieving similar results will be quite challenging due to the persistent problem of radioactivity of materials. Even now, Borexino remains a unique detector in the world for its radiopurity.

There is another important legacy of this experiment, which lies in the way we proceeded—challenging the skepticism of the scientific community that surrounded us, given the difficulty of the experiment. Many colleagues advised against moving forward. Yet, I am convinced that without taking risks, it is impossible to achieve truly significant goals, and this is especially true in physics, where the most important results often come from undertaking very difficult experiments. These considerations are crucial for young people, for whom the risk may seem even greater at the start of their careers. As I have mentioned earlier in this book, I always felt a sense of responsibility for encouraging young colleagues to dedicate themselves to this experiment, driven by the fear that it might not succeed. But I have always found within myself the determination to press on and the confidence to inspire my colleagues.

Now that everything has gone well and the experiment has exceeded initial expectations, I feel relieved of that responsibility. I would like to conclude by recalling a quote from John Bahcall, the father of the standard solar model, a great friend, and a supporter of the experiment: "The most important discoveries will provide answers to questions that we do not yet know how to ask and will concern objects we have not yet imagined."

Glossary

Anode In electronic devices, a suitably shaped metal object that represents the positive pole, which normally has a higher potential.

Best fit In data analysis, particularly in the context of plots derived from physics experiments, the term refers to the curve that most accurately approximates a distribution of points, such as those resulting from measurements. It can also refer to the function (best-fit function) that provides the best analytical approximation of the distribution.

Cherenkov effect It is the phenomenon by which an electrically charged particle, whose speed is a relevant fraction of the speed of light, as it passes through matter produces photons of light, with a direction following that of the incident particle.

Clean room Clean rooms are controlled environments with regulated air quality. The air inside is filtered through resin filters capable of capturing particles at the micron level, ensuring it is cleaner than external air. For certain specialized clean rooms, radon gas is also removed using cryogenic methods.

The classification of clean rooms is based on the number of particles measuring ≥0.5 μm in size. Classes are defined as follows: Class 10, Class 100, Class 1000, and Class 10,000, indicating that the number of particles ≥0.5 μm must not exceed 10, 100, 1000, and 10,000 per cubic foot, respectively.

The measurement unit "cubic foot" is used because clean rooms were first introduced in the United States. One cubic foot corresponds to approximately 0.03 m3. Work within clean rooms must be performed with great care to avoid contaminating the air. In particular, operators are required to wear specialized suits, including masks and hair covers. One of the main sources of air contamination has been identified as human breath.

Glossary

Decay Radioactive decay is the transformation of a nuclide into another nuclide, which can also be either radioactive or stable. This decay occurs after a certain characteristic time for each nuclide.

Doppler effect The Doppler effect is a physical phenomenon that consists of the apparent change, relative to the original value, in the frequency perceived by an observer reached by a wave emitted by a source in motion relative to the observer. The Doppler effect is particularly important in the study of the motion of stars, which, if moving away from us, exhibit a shift of light toward the red end of the spectrum. This effect is known as "redshift".

eV-electronVolt Specialized units, which represent the energy gained by an electron moving through an electric field with a potential difference of 1V. Since 1 eV corresponds to a very small amount of energy—1.6×10^{-19} Joule, the unit of energy in the international system—larger multiples such as kiloelectron-volts (keV), megaelectron-volts (MeV), and gigaelectron-volts (GeV) are often used in physics. It is particularly used in nuclear and subnuclear physics, specifically in the study of elementary particles.

Flavor In elementary particle physics, the term flavor refers to distinct properties that differentiate types of a given particle. For example, in the case of neutrinos, flavor distinguishes between the different types of neutrinos: electron neutrino, muon neutrino, and tau neutrino.

Ion An ion refers to an electrically charged atom, typically resulting from the loss or gain of one or more electrons. The process by which an atom loses or gains electrons is known as ionization.

Isotank It is a specialized container designed for transporting liquid or gaseous products such as fuel, chemicals, cement, or tar. To ensure safety and functionality, it must comply with specific requirements, including a minimum wall thickness, the ability to withstand internal pressurization, maximum sealing efficiency and adherence to various safety and performance standards.

Mass spectrometer The mass spectrometer is an analytical instrument used to separate ions with the same charge but different masses, or, more generally, ions with differing mass-to-charge ratios, such as isotopes.

Muons (μ particles) Elementary particles belonging to the lepton group.

Neutrino It is an elementary particle that belongs to one of the two families of them. The family it belongs to is composed of the so-called leptons, derived from the Greek word "leptòs," meaning light. This family also includes the electron. There are three different types of neutrinos: electron neutrino, muon neutrino, and tau neutrino.

Neutrino Neutrinos are so small and light that their size and mass have not yet been measured, as current technologies are not capable of detecting such tiny values. They possess an incredible and highly useful property: they interact with matter only very much weakly.

Neutron It is one of the two particles that make up the atomic nucleus, also known as nucleons. The neutron is electrically neutral.

Nuclear fusion It is the combination of two light nuclei to form a heavier one. The resulting nucleus has a mass slightly less than the sum of the masses of the original nuclei. For example, when four hydrogen nuclei fuse to form a helium nucleus, the helium's mass is less than the combined masses of the four hydrogen nuclei. This "missing" mass is converted into energy, which is released.

Nuclide The term *nuclide* refers to an atom characterized by a specific number of protons in its nucleus and a specific number of neutrons. Unless otherwise indicated, the term generally implies that the atom is electrically neutral, meaning the number of electrons surrounding the nucleus is equal to the atomic number.

Photomultiplier An electronic eye that captures the photons produced by the scintillator, converts them into electrons, multiplies them, then transforms into an electric signal.

Photon A quantum of energy carried by an electromagnetic wave when it is part of the light spectrum.

Proton It is one of the two particles that make up the atomic nucleus, also known as nucleons. The proton is electrically charged in a positive way.

Radon-222 Radon, specifically Radon-222, is a radioactive gas that belongs to the natural radioactive decay series originating from Uranium-238. Radon is produced through the decay of Radium, which is its parent in the radioactive family.

Since Radium, albeit in small quantities, is present everywhere—in materials, rocks, and so on—Radon is ubiquitous, especially in poorly ventilated areas where it cannot disperse.

Radon decays by emitting what is known as an alpha particle, which is essentially a Helium nucleus. Alpha particles are easily absorbed, particularly by human skin, and are therefore not dangerous unless inhaled.

Scintillator A liquid or solid material that emits photons of light in all directions when traversed by a charged particle that releases energy within it.

Spectroscopy It is referred to by this name the decomposition of light into its component wavelengths, achieved by passing it through a dispersive element such as a prism, transparent optical device, typically made of glass or crystalline materials.

This occurs because the refractive index of the prism's material varies with the wavelength of light: higher frequencies (such as blue and violet) are refracted more, while lower frequencies (such as red) are refracted less. This effect makes prisms essential tools for the spectroscopic analysis of light.

Each atom, when excited by receiving energy, emits light of various colors corresponding to different frequencies of the light wave. These frequencies are characteristic of a specific atom, and thus, by observing the various emitted colors (and therefore the different frequencies), it is possible to identify the atom in question. In other words, the spectrum of an atom corresponds to its fingerprint.

Spectrum The optical spectrum refers to the decomposition of light into its components. This means that light can be the combination of various colors, as seen in white light, which is the combination of all possible colors. Since light is nothing but an electromagnetic wave, it can take different forms: in the optical range with various colors corresponding to different frequencies of the waves. The optical spectrum is obtained by passing light, for example through a prism, made of glass or crystalline materials, which breaks it down into its various components.

Threshold It generally refers to something above which an event materializes. In physics experiments and in reference to energy, it means the energy value above which measurements can be made, while below this value instrumental limitations prevent measurement.

The manufacturer's authorised representative in the EU is Springer Nature Customer Service Centre GmbH, Europaplatz 3, 69115 Heidelberg, Germany. If you have any concerns regarding our products, please contact ProductSafety@springernature.com

Printed and bound by CPI Group (UK) Ltd, Croydon, CR0 4YY

26/03/2026

02078940-0014